Applied Mathematics
+ Operational Research
Cranfield University
RMCS Shrivenham
SN6 8LA
Tel: 01793 785294

Modeling and Simulation in Science, Engineering and Technology

Brunello Tirozzi
Silvia Puca
Stefano Pittalis
Antonello Bruschi
Sara Morucci
Enrico Ferraro
Stefano Corsini

Neural Networks and Sea Time Series

Reconstruction and Extreme-Event Analysis

Birkhäuser
Boston • Basel • Berlin

B. Tirozzi, S. Pittalis, A. Bruschi,
S. Morucci, E. Ferraro
Dipartimento di Fisica
Università di Roma "La Sapienza"
2 Piazzale Aldo Moro
Roma 00185
Italy

S. Puca
European Meterological Satellite
Organization (EUMETSAT) at the
Italian Meteorological Institute
of the Air Force
Rome 00040
Italy

S. Corsini
APAT (Agenzia per la Protezione
dell'Ambiente e per i Servizi Tecnici)
3 Via Curtatone
Roma 00185
Italy

AMS Subject Classification: 05D40, 37M10, 60G25, 60G70, 62M45, 82C32, 92B20

Library of Congress Cataloging-in-Publication Data
Neural networks and sea time series : reconstruction and extreme event analysis / Brunello Tirozzi . . . [et al.].
p. cm. – (Modeling and simulation in science, engineering & technology)
Includes bibliographical references and index.
ISBN 0-8176-4347-8 (alk. paper)
1. Oceanography–Statistical methods. 2. Times-series analysis. 3. Neural networks (Computer science) I. Tirozzi, Brunello. II. Series.

GC10.4.S7N48 2005
551.46′072′7–dc22 2005043635

ISBN 10-8176-4347-8 eISBN 0-8176-4459-8
ISBN-13 978-0-8176-4347-8

Printed on acid-free paper.

Printed in the United States of America. (TXQ/IBT)

9 8 7 6 5 4 3 2 1

www.birkhauser.com

Contents

Preface

This book describes the results of a successful collaboration between a group at the Department of Physics of Rome University "La Sapienza" and a group of engineers who for many years have been engaged in researching marine events at the National Department of Technical Services of Italy ("Dipartimento dei Servizi Tecnici Nazionali, DSTN, now APAT, Agency of Environmental Protection and Technical Services). From the time when mankind first took to the sea, the possibility of creating safe and secure marine constructions has given rise to the need to monitor the complex phenomenology of the sea's evolution. Our group of physicists has a lot of experience in the utilization of artificial neural networks (ANN) in many different fields and has always had the feeling that many improvements could result from their applications. On the other hand, the engineers have a specific problem in dealing with the long time series of sea data, which are necessary for security estimates and improving the applications. The exchange among the two groups led to the conclusion that this problem, together with many others described in this book, could be solved by the application of ANN.

The particular problem that we take as our starting point is the reconstruction of missing data, a general problem that appears in many situations of data analysis, although the solution obtained with the application of ANN was reasonably good, albeit unexpected. This fruitful attempt is one of the first applications of ANN to time series of sea data that we know of. We hope that this promising beginning will encourage researchers in the field to continue to develop such an approach to similar problems.

Since this is one of the first experiences in this field, we feel that it is useful and important to describe in detail how one can use ANN, a flexible but complex instrument, and also to explain the theory behind it. At the same time it is necessary to show the complex problems posed by sea events and their phenomenology in order to understand exactly what we have to investigate. One of the main steps in the application of ANN is to have a sure sense of the problem to be dealt with. Otherwise, even a good technical knowledge of the algorithm is useless. This book therefore includes both the theory and practice involving these two fields. It is written in such a way that nonexperts in both fields can understand it. We hope that this presentation may encourage readers to try an analogous approach in other useful applications. In particular, we want to emphasize that the study on the sea time series measured along the Italian coasts presented in this book can be easily repeated and generalized to other coasts and seas. We give an example of this in Chapter 10, where we analyze the same problems and solutions relative to the sea heights and levels around the coast of California.

We thank Professor Sergio Albeverio and Professor Antonio Speranza for many useful discussions and their careful reading of the manuscript.

Brunello Tirozzi
Silvia Puca
Stefano Pittalis
Antonello Bruschi
Sara Morucci
Enrico Ferraro
Stefano Corsini

Neural Networks
and
Sea Time Series

1

Introduction

1.1 General Remarks

There are some important reasons for writing this book about neural networks (NNs) even though many books and articles on this topic have already been published. One reason is that we deal with the application of NNs as an algorithm for data analysis in the field of sea phenomena. This is quite a new field of application for the NN; few papers have yet been published using this approach. NNs have a double significance: one is that they provide very flexible and adaptable algorithms to handle data analysis and the other is that some versions of these algorithms can be used to describe the behavior of real neurons. The first meaning of NN is the one which is dealt with in this book. We do not deal with all the possible applications to the complex world of time series, which is impossible to do in the space of one book (it would require an entire encyclopedia), but only with the applications of NNs to time series of sea levels and other related important variables.

The field of sea waves is very important and has been studied and developed for a century or more. It is studied intensively for climate studies and is also the center of attention for marine constructions; e.g., ports and oil platforms. However there are still open problems in the theory describing these important phenomena. The kinetic equations describing the evolution of the sea and oceans do not have a rigorous derivation. The determination of the wind field cannot be done over the entire domain of the application of the equations and so some extrapolation technique from the measurement points must be applied, thus leading to errors in execution. The errors in the prediction and reconstruction based on these models have been improved by use of the postprocessing technique, something we discuss in this book. The dissipation terms appearing in the governing equations do not have a good validation. There are many alternative proposals to the equations (see, for example, the papers of Onorato et al. [2], Dysthe and Trulsen [3], Zakharov [4]) and different estimates of energy spectrum of the waves as shown in Pelinovsky [5]. Thus, a deterministic approach still needs some serious work to arrive at a completely satisfactory theory. These difficulties and the necessity of having better local forecasts of the sea parameters have convinced some researchers to use statistical methods for treating these particular phenomena. NNs can be considered as a powerful new statistical tool in the sense that they can be used for sea state local predictions and reconstructions of gaps in a time series. They have been used in the field of sea waves and sea levels only quite recently and there are still few publications on this subject ([1]). The usual statistical methods such as Kalman filters and ARMA (autoregressive moving average) models are most commonly used, but in this book we show that NNs more often get smaller errors in the approximation of data. We will compare the results of ARMA for the reconstruction of time series with the results obtained by NN and show that NNs give a better answer. Furthermore, we show an application of NN to a related field, the postprocessing of temperatures, and compare the NN results with those of the Kalman filter, again showing more efficient results.

This book can also be considered as an efficient example of how it is possible to apply the NN to a large set of data. It is comprehensive in the sense that many aspects of the theory underlying the application of NNs are explained, trying to fill a gap in the applied works concerning the use of the NN. The reader of such papers could get the impression that NNs are tools to be used in almost any situation of data analysis without any background on theoretical aspects. This impression comes from the fact that the theoretical and the applied literature are usually distinct in many fields of applied mathematics and engineering. Theoretical results follow many different directions and we try to highlight some of them by explaining the probabilistic aspects and the meaning of the NN as a universal system of complete functions (Chapters 4 and 5). We also report and utilize useful theories and mathematical questions treated in other books. We believe that collecting theories or experimental facts published elsewhere is useful in order to produce a self-contained book. Also, for ease of reading, we extract from the articles and books quoted in the References the necessary information and tools. Some theorems and theories are only partially used in the applications described in the final chapters but we want to provide a wide enough theoretical background for possible generalizations. In Chapter 4 we use the term *artificial neural network* (*ANN*), which is currently used for an NN applied to time series. We also explain the problems from the general application of NNs, i.e., the generalization error and learning error (Sections 4.3 and 4.4). Our coverage of this wide range of theoretical and applied aspects will be generally useful for people who need to use NNs for specific problems and require some background knowledge.

This book is also complete in the sense that almost all the important issues concerning sea levels and wave phenomenology are treated. We also explain extreme-event analysis (Chapter 6) and the problems and the measurements of wave heights (Chapter 2). We present the results of the application of extreme-event algorithms to the time series generated by the system of measuring buoys located along the Italian coast (Chapter 9). In Chapter 10 we also study the case of other sea time series and of other important physical variables. In this chapter we also discuss the application of NN to the time series generated by the buoys located near the California coast. In the same chapter we show the use of an NN for improving temperature forecasts, in a certain station of southern Italy and the NN's application to the prediction of precipitation (postprocessing techniques). In Chapter 11 we present some conclusions and perspectives for future work. The first six chapters of this book contain all the information and theories used in the later chapters. The last chapter contains our conclusions.

1.2 Plan of the Book

As already noted, there are many different aspects analyzed in this text, which is due to the fact that there are many different experimental and theoretical methods for the analysis of the evolution of the sea. The complexity of the book's contents derives from our attempt to show NN theory in all its aspects. In Chapter 2 we give some general information on the phenomenology of waves and tides and the related technical details of the various measuring processes. We also discuss the importance of understanding waves and tide distribution for marine engineering activities related to the sea. We also describe the network of buoys located around the Italian coast and its function. This network has provided all the time series on which we used our algorithms. There is also a definition of what is meant by wave height by the operational level, a description of the various components which contribute to tide height and dynamics as well as a discussion of how tide phenomena are affected by the planets. This chapter is useful for understanding the physical quantities that are used throughout the book. In Chapter 3 we describe the model of wind waves (WAM, the wave model). This model is used in Europe to determine the spectral function of the waves and for predicting the behavior of the SWH (significant wave height). We compare the results of a reconstruction made by using the NN algorithms of the SWH time series with the SWH computed by the WAM in the point nearest to the measuring buoy. We made this comparison in order to check the order of magnitudes of the SWH evaluated with the NN. For the same reason we compared the reconstructed time series of tides and sea levels (SL), with those evaluated by means of an astronomical model used in many centers of operation. In Chapter 4 we describe all the necessary information for the use of ANN (Sections 4.1–4.3). The more theoretical considerations are given in Section 4.4. In Chapter 5 we explain the meaning of the NN as a complete set of functions and how one can use it in this sense. In Chapter 6 we explain our main methods and ideas for analyzing the probability distribution of extreme events. We first discuss the case of independent equally distributed random variables and the method called peak over threshold (POT) that is largely used in marine applications. We also discuss the case of dependent random variables, even if it is not used in the applications, since it is useful to know this for generalizations. Chapter 7 discusses the main application of this book. There we show the reconstruction of significant wave heights (SWH) and sea levels (SL) using ANN. We also show how we implemented an automatic detection system and filled in the gap in the data time series (Section 7.4). In Chapter 8 we analyze (Section 8.0) the application of ANNs used as approximation operators, in the sense explained in Chapter 5, to the real data and show why it is better to use ANNs rather than the approximation operator approach. In Section 8.1 we show the limits of the use of the autoregressive integrated moving average (ARIMA) models for the SWH. In Chapter 9 we show the results of extreme-event analysis (POT method) applied to the time series of several stations (Sections 9.1–9.3). There is also a description of a special ANN used to treat the large values of the heights and the connection among the waves reconstructed ANNs and the statistics of the extremes (Section 9.4). In Chapter 10 we discuss how to treat analogous problems for a system of buoys along the California coast. In the same chapter we also show how ANNs are successful for other types of data and problems. We discuss the temperatures measured in some stations and how one can use them for predictions of the future values by using the estimated values of the global model used at the ECMWF (the European Center for Medium Range Weather Forecasts). In other words, we describe how to improve the temperature forecast of the meteorological model of the ECMWF using the data measured by a station; this procedure is called *postprocessing*. We also discuss how the forecast of the precipitations can be improved using NN and obtaining predictions similar to or sometimes better than the ones of the meteorological models. We remark that the language of the book is not homogeneous because Chapters 4, 5, and 6 on the methods and models are written in the usual mathematical language which differs from that used in the more phenomenological chapters. This discrepancy is due to the intrinsic difference between the two topics and cannot be avoided (unfortunately) since we want to give a full description of both. In the phenomenological chapters the concepts and the definitions of the theoretical chapters are widely applied and explained so that readers can understand their applications to a concrete case. Thus, they can first read the applications and then consult the theoretical chapters for more exact definitions and explanations or skip the mathematical chapters entirely.

1.3 Introduction to Waves

The open sea is under the constant action of external natural forces which generate various types of waves. The most common cause of the development of surface waves of the sea is the action of the wind. The ancient Greeks were already well aware of the phenomenology relating to the interaction between the atmosphere and the sea surface. Aristotle understood that the wind flowing on the sea surface plays a fundamental role in wave development. Pliny made the remark that dropping oil on the sea reduces the motion of the capillary waves. From the time of Aristotle until the Renaissance and the Golden Age of Discoverys at the end of the 15th century, very little progress was made toward understanding the mechanism governing the sea waves. Only in the 19th and in the 20th centuries was a clearer understanding reached concerning the fundamental principles governing the appearance of the waves on an undisturbed surface and the determination of the waves' evolution in space and time.

In general, we can identify five main wave types in the open sea: sound, capillary, gravity, internal and planetary. The sound waves are due to the compressibility of the water, which is rather small. Gravitational forces acting on the fluid particles deviating from the equilibrium position on the sea surface or on a geopotential surface internal to a stratified fluid determine the gravity waves (internal or surface waves). There are other types of internal waves but we do not discuss them. At the air-water interface the combined effects of the turbulent action of the wind and the surface ten-

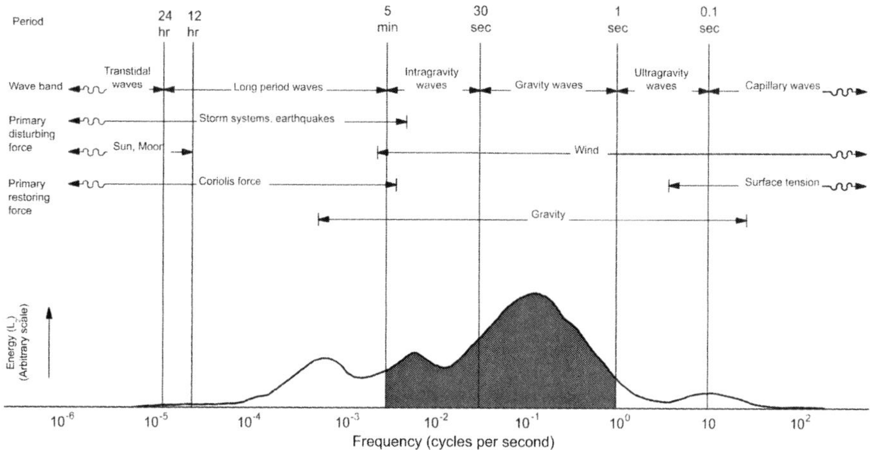

Fig. 1.1. Schematic representation of the energies contained in the surface waves.

sion give rise to capillary waves with short wavelength and high frequency. On the other hand, planetary waves on a large scale (planetary or Rossby waves) are produced by fluctuations of the potential vorticity equilibrium values due to a change of depth or latitude. All these kinds of waves can form at the same time giving rise to complicated oscillation patterns. The external forces are distributed along a wide frequency interval and the corresponding motions of the sea surfaces range in an exceptionally wide field of amplitudes and wave periods. From capillary waves, with a period of less than a second, one can observe wind and sea-swell waves with periods up to less than few seconds or tens of seconds to tide oscillations with periods of hours or days. Fig 1.1 and Table 1.1 provide a schematic representation of the energies contained in the surface waves and a list of the possible physical mechanisms which generate them.

Table 1.1. Waves, physical mechanisms and periods.

Wave type	Physical mechanism	Periods
Capillary waves	Surface tension	$<10^{-1}$ s
Wind waves	Wind shear, gravity	<15 s
Swell	Wind waves	<30 s
Surf beat	Wave groups	1–5 min
Seiche	Wind variation	2–40 min
Harbor resonance	Surf beat	2–40 min
Tsunami	Earthquake	10 min–2 h
Storm surges	Wind stresses and atmospheric pressure variation	1–3 days
Tides	Gravitational action of the moon and sun, earth rotation	12–24 h

From Fig 1.1 it is possible to get an idea of the relative importance of the various kind of sea surface oscillations. The indicated periods are just orders of magnitudes, as, for example, in the case of a tsunami where there can also be periods longer than 2 hours. For engineering activity at sea, the gravitational waves are the most important since the action of the waves generated by the wind on the sea structures is the strongest and the most adverse. Substantial interest in the most significant contributions to the knowledge of the mechanics of wind waves began almost one century ago. The main contributions were analysis of the linear theory of the waves (Airy, 1825), the higher order theories (Stokes, 1847), the theories of long waves (Boussinesque, 1872) and the theories of maximum wave height (Michell, 1893 and McCowan, 1894). After half a century of pioneering developments, the research on this topic continued at a relatively slow pace until the amphibious landings of World War II. These enhanced the need for a better comprehension of the wave motion due to the wind. Also the conservative and dissipative mechanisms occurring in the generating areas and the approach to shallow coastal depths and the process of waves breaking on the beaches became important questions The great failure of using floating and removable breakwaters during the landing in Normandy stimulated the interest in a deeper study of the interaction between waves and various structures. After World War II research on waves would probably have been neglected without the explosive growth of maritime engineering activity in the scientific, industrial and military sectors. From 1950 to1990 sea drilling and the production of oil resources moved from shallow depths of 10 meters to much greater depths of 300 meters. As a result, oil platforms were planned for wave heights of the order of 25 meters and costs in the range of hundreds of millions of euros. The financial incentives for well-planned and realized studies of sea wave phenomena increased accordingly. Laboratory investigations as well as field measurements (much more expensive) were required for to validate design methodologies and obtain a better knowledge in order to describe the complex characteristics and non-linearities of multidirectional sea states. A second and more substantial impulse for sea wave research came from interest in the phenomena of coastal erosion, until recently a field explored only superficially. A complete understanding of the interaction of the waves with structures of the sea and with the morphologic and sedimentary structure of the coast is crucial for the activities of the men at sea. The computational schedule necessary for establishing the structural load on sea constructions is made up of the following steps, in each of which a thorough knowledge of sea surface waves is essential:

- Definition of the wave climate in the area of interest
- Estimation of the characteristics of wave motion
- Selection and determination of the hydrodynamic load on structures.

The role of waves in environmental matters cannot be underestimated. Waves propagating toward the coast break and dissipate their energies on the beach, conditioning its equilibrium. The big waves of a storm act with enormous forces on both the natural and artificial coasts. The coastal currents generated by the wave motion transport large quantities of sediments, creating waves of accumulation and erosion.

A knowledge of the characteristics of the wave motions and of the sedimentary budget makes it possible to choose an appropriate protective structure.

1.4 Introduction to Tides

Tides are oscillations of the sea level determined by the action of astronomical forces. When an astronomical body is under the gravitational influence of one or more masses, then tide forces act on its liquid components. These forces are the sum of the centrifugal force caused by the revolution of the celestial body around the common mass center of the system and of the gravity attraction of the nearest bodies. The earth belongs to such a system and the tide forces acting on it are essentially due to the moon and to the sun. Due to the features of the relative motions of the sun and the moon with respect to the earth, such forces have mainly daily and half-daily periods, even though there are many other characteristic periods.

Since antiquity man has tried to make forecasts of the tides, but only at the end of the 1700s were quantitative attempts made by Laplace (1775) and later by Kelvin and Roberts (1868). Any attempt at tide forecasting requires a careful analysis of the experimental data which essentially consist of the measures of the elevation of the sea level repeated on more or less long time intervals. Primitive analysis consisted of correlating, at least qualitatively, the observations with astronomical events, and the success of such an approach seemed to be the result of magic powers. Progress in other fields of applied mathematics showed later that the properties of astronomical tides can be obtained by general time-series analysis of the SL and of the wind-generated components of long waves. This method allows a strong synthesis of the information obtained from the observed data. The information contained in a paper coil reporting the continuous signal of the sea level in a station during one year (a sheet of paper about 90 meters long) may be condensed into a few values which can be very effective in forecasting the sea level's future behavior caused by planetary influence. In fact, the regularity of the astronomical motions implies the presence of periodic components in the time series of the tide. Indeed, to research such periods is the main aim of this analysis. A series of level measures at a given point on the earth's surface can be approximated using a sum of harmonic terms. The fit turns out to be very good, and significant differences between the series of the measured signal and the series reconstructed with the harmonic terms can only be explained by considering phenomena independent of the astronomical tide. During the 19th century various analytical tools were developed; these aimed at finding the amplitude and the phase of the principal harmonic components (frequencies) of the measured data, removing the fluctuations due to non-astronomical effects. The choice among different mathematically correct approaches is almost irrelevant once the principles governing the phenomenology are universally accepted, as they effectively are nowadays.

2

Basic Notions on Waves and Tides

The characteristics of wave motion are important for understanding the evolution of a coast. In the first three sections of this chapter we give the definitions of the quantities describing waves and a description of the current instruments and methodologies for their measurement. We describe the network of buoys used for attaining the significant wave height (SWH) time series analyzed in this book. We use a similar approach for tides: some of the theory and the corresponding network of buoys are described.

2.1 Definitions of SWH

The buoys of the network measure four main parameters associated with the wave motion. We describe them in this section. The main parameter is the directional spectrum. Let $\eta(t)$ be the elevation of the sea surface that can then be expressed as the sum of different waves with frequency f_i and wave vector $\mathbf{k}_i = k_i(\cos\theta_i, \sin\theta_i)$:

$$\eta(t) = \sum_i a_i \sin(k_i(\cos\theta_i x + \sin\theta_i y) - f_i t + \phi_i). \tag{2.1}$$

The vector $\mathbf{k}_i$ defines the propagation direction of the wave and ϕ_i is a random variable associated to the randomness of the wave. The spectral directional density is defined by the sum

$$E(f, \theta) = \sum_i \frac{1}{2} a_i^2, \tag{2.2}$$

where the sum is calculated over all i such that $\omega_i \in (\theta, \theta+d\theta)$ and $\omega_i \in (f, f+df)$. We define the normalized density as

$$D_f(\theta) = E(f, \theta)/E(f), \tag{2.3}$$

where $E(f, \theta)$ is the spectral directional density and $E(f) = \int_0^{2\pi} E(f, \theta)d\theta$ is the energy density.

We define the SWH H_s, the significant wave height, T_m the average peak period and T_p the peak period using the energy density. Let q_i be the spectral moment of ith order

$$q_k = \int_0^\infty f^k E(f)df; \tag{2.4}$$

then the SWH H_s is

$$H_s = 4\{q_o\}^{1/2} \tag{2.5}$$

and the T_m is

$$T_m = q_{-1}/q_o. \tag{2.6}$$

The value of T_p is the inverse of the frequency f in which $E(f)$ takes on the maximum value. Together with the average directions of the waves these are the four parameters transmitted by the buoys to a data server. The classical definition of SWH, equivalent to the preceding definition of H_s, is given by using the dispersion of η. Let us show this alternative definition for the sake of clarity and thoroughness. The energy density $E(f)$ is equal to the Fourier transform of the autocorrelation function of the wave height $\eta(t)$. The exact definition is

$$\Phi_{\eta\eta}(f) = E\left|\frac{1}{T}\sum_{t=0}^{T}\eta(t)\exp(2\pi i f t)\right|^2, \tag{2.7}$$

where E is the mean with respect to the stochastic process associated to the $\eta(t)$, the sum over t being the approximation of the integral with the integral sums. Making the square and doing the expectation we obtain

$$\Phi_{\eta\eta}(f) = \frac{1}{T}\sum_{\tau=-T}^{T} C(\tau)\exp(2\pi i f \tau), \tag{2.8}$$

where $C(\tau) = E\eta(0)\eta(\tau)$ is the autocorrelation function of the wave height and $\Phi_{\eta\eta}(f)$ its Fourier transform. Using the autocorrelation function it is possible to show the equivalence with the more classical definition of SWH given in terms of the time average of the square of the wave heights:

$$\sigma^2 = \int_0^T \eta(t)^2 dt. \tag{2.9}$$

Actually, the real classical definition of SWH was given considering wave heights observed in the time interval $(0, T)$, $\eta_1, \ldots, \eta_n$ and ordering them taking η_1 to be the maximum, η_2 the second-order maximum and so on. Then the SWH H_s is

$$H_s = \frac{\eta_1 + \eta_2 + \cdots \eta_{n/3}}{n/3}. \tag{2.10}$$

Using the law of large numbers or the ergodicity of the stochastic process describing the wave motion, it is easy to connect the two definitions:

$$\sigma^2 \sim E\sigma^2 = C(0) = \int \Phi_{\eta\eta}(f)df = \int E(f)df. \tag{2.11}$$

In other words, we can say that the square of the SWH H_s defined in (2.10) as a certain time average of the η_i coincides, for enough long observations, with the square of the SWH defined in (2.5) by means of the energy density.

2.2 Experimental Apparatus

Ultrasound sensors

Ultrasound sensors, as in the case of tide gauges, produce a packet of ultra-sounds and receive an echo reflected from the sea-air interface allowing a measure of the distance between the transducer and the instantaneous surface of the sea. The typical emission frequencies range from 10 kHz for the air measure (with the sensor mounted on a fixed support at a level higher than the maximum height to be measured) to 200 kHz for the underwater case with an emission cone of about 10°. It is necessary to use a calibration system for taking into account the salt and temperature variations.

Radar sensors

Radar waves with typical frequencies around 10 GHz can also be used for measuring the distance between the level of the sensor, placed on a platform, or a pole, a bridge or a ship and the sea surface. One of the most modern sensors is very small and is based on the elaboration of a digital signal and a plane antenna with a measure field up to a distance of 65 meters. The entire flat area of the antenna emits and receives a frequency spectrum with an angle of 5°. The radar sensors are highly accurate (± 1 mm with respect to the rest surface).

Pressure sensors

The measure of the wave motion by means of the fluctuation of the hydrostatical pressure induced by the waves is limited to the shallow water area since the pressure fluctuations of the highest frequency on a dipped sensor decrease rapidly with increasing depth. This implies that short period waves of the sea state are filtered in the measuring process and that the average period T_m results are higher than the real ones. However, these sensors have been shown to be robust and cheap and capable of measuring the wave height with a precision of 10%. With a correct calibration of the transfer function they are then able to measure the long-term wave motion in the coastal area, particularly for the measure of the long-period waves (e.g., seiche, setdown and surfbit components).

Level sensors coupled with current meters

Level sensors may be coupled with a double axial current meter without internal independent parts which measure the components of the current velocity along the two main directions at a certain level obviously synchronized with the measure of the change of the surface level. The direction of the wave motion is computed on the basis of the north and east components of the orbital velocity of the waves. Such a system can only be applied to depths that are not to great (maximum 20 m) because of the quick reduction of the orbital velocity with increasing depth. The instruments equipped with pressure sensors can be placed on the seabed, at the top of a protected structure (metallic tripod, conic or pyramidal armor) opportunely blocked at the base. For coupling the current meter with an ultrasound or radar sensor, a robust emergent support is necessary.

Directional wave buoys

The directional buoys can be grouped into two main classes according to the principles of measurement: the "slope-following" type and the translation type. The Wavec buoy, produced by Datawell (2.5 m in diameter and 750 kg weight) and the Wavescan buoy of the Oceanor (2.8 m in diameter and 900 kg weight) belong to the first class. This kind of buoy follows the movement of the sea surface and measures wave height and direction with respect to an inertial platform, contained inside the sensor, which corresponds to the reference horizontal plane. The vertical displacement is obtained by a double integration of the vertical acceleration measured by an accelerometer mounted on the platform. The inclinations are derived from the measure of the sine of the angles between the x and y axes of the reference system in strict connection to the buoy and the reference horizontal plane. A system of compasses measures the components of the magnetic field: H_x, H_y, H_z. From these components of the magnetic field and from the preceding inclinations it is possible to derive the following: the absolute value and inclination of the earth's magnetic field, the direction of the buoy and the two local slopes of the water surface along the north-south and east-west directions, thus making it possible to define the direction of the waves' propagation. The Waverider buoys produced by Datawell (0.9 m in diameter and 200 kg weight) and the Triaxis buoy produced by Axis are of the second type. The difference from the buoys of the slope following type is that the sensor measures the shifts with respect to the horizontal plane instead of measuring the slopes with respect to the principal axes. This translation system is more efficient in terms of data transmission and provides ease of installation and maintenance.

2.3 The National Sea Wave Measurement Network (RON)

Since the end of the 1970s more than 100 measuring stations managed by 25 different institutions have been operating in Italy in different periods. Often the information

Fig. 2.1. RON.

on particular projects came from fragmentary time series and was obtained with big-type instruments, most of them buoys. Since July 1989 the Italian *Rete Ondametrica Nazionale* (RON) (National Sea Wave Measurement Network) Has been active. RON is made up of 14 directional buoys, located as shown in Fig. 2.1, off the coast of La Spezia, Alghero, Ortona, Ponza, Monopoli, Crotone, Catania, Mazara, Cetraro, Civitavecchia, Capo Cimino (Sardinia coast), Palermo and Punta della Maestra (outfall of Po River). Each buoy, berthed at sounding depths of about 100 meters, follows the movement of the surface of the water and measures the waves' height and direction. The instruments are equipped with a localization system which uses a satellite to continually check position.

During these years RON has maintained a very high-quality standard with respect to the efficiency of the instruments, especially since 1996, following the new demands and perspectives determined by the rapid developments in computer science and communications. The management of RON by the Marine Service of APAT (Agenzia per la Protezione dell' Ambiente e dei Servizi Tecnici also called Agency of Environmental Protection and Technical Services has made it possible plan and carry out a real-time data elaboration system which is today integrated with a new and efficient real-time network of tide gauges called RMN (the new Italian *Rete Mareografica Nazionale*, or National Sea Level Measurement Network). Taken as a whole, such a system is at present the most complete tool of analysis in real time of the sea state in the Mediterranean, including continuous monitoring of the sea for the different possible phenomenologies, from tides to storms, as well as methodological information near the coast. Regarding the activities developed in the 15 years of RON, the main innovations are the realization of a transmission system and storage of all the parameters obtained from the buoys (including both spectral analysis and the analysis of the time series) and the visualization in real time of the measures on an internet site dedicated to RON. All the changes have been made keeping the instrumental characteristics and measure standard defined at the moment of the installation of the network. This attempt at technological updating and continuity of the measurement process has resulted in the availability of time series unique in the Mediterranean area for high quality and continuity. Another contribution has been made in making it possible to utilize the information given by the RON to produce a series of statistical analyses concerning the network's functioning during the entire period under consideration. The elements relative to the sites' climatologic conditions have been incorporated in the same time series. This information is the result of monitoring and research developed by the Marine Service of APAT in this period. These data constitute the basis for defining the sea state by means of tables and diagrams of the significant wave height, incoming average direction, peak period and average peak of the wave events. Using such climatologic considerations as a starting point, a statistical analysis of the return times of the extreme events has been carried out. This contribution is not considered as a total explanation of the elaboration of the available data, but as a starting point for the improving knowledge of our sea climate. At present new studies are being developed on the following topics:

- The reconstruction of the missing extreme wave events (for random wrong functioning of the instruments, for unmooring of buoys and for transmission problems) both by means of high spatial resolution numerical models and neural networks
- The publication of a catalog of extreme events measured by RON with an analysis of the meteorological extreme events associated with them.

Technical characteristics of RON

Fourteen Triaxis buoys belong to RON. Data are transmitted continuously via radio waves on the 44.8 MHz band to a receiver located on shore. They are received normally for a period of 26 minutes every half hour. By means of the procedure described above, the following quantities are obtained:

1. The synthetic parameters
    - H_s, significative wave height
    - T_p, peak period
    - T_m, mean period
    - D_m, mean incoming wave direction

2. The spectral parameters for each frequency band
   - energy density
   - propagation direction
   - directional spreading
   - skewness
   - kurtosis

In the onshore stations there are two computers, connected with the receiver. These receive the raw and elaborated data transmitted from the buoys and transmit a synthesis of the data to the Center of Control and Management of the network situated at the headquarters of APAT's Marine Service.

RON's wave data

The sample of the measured data is filtered on the buoy with a Butterworth low passing filter of the seventh order to prevent errors due to aliasing. Vertical, east, and north displacement data are sampled every 0.25 s, and processed and transmitted onshore every half hour in a binary file decoded by the receiver. Transmission errors are found and corrected.

2.4 Tides

In this section and in the two following ones we give a survey of the main problems and the most commonly used methods of measurement connected with sea level and tides. The tide force at a certain point P on the surface of a given planet is the sum of the Newtonian attraction of another planet (sun or moon) and the centrifugal force of the revolution motion at the point P. The force of gravity at the same point is nothing more than the sum of the Newtonian action of the earth (with the mass concentrated in the center) and the centrifugal force generated by the planet rotation around its own axis. The tidal force deriving from a very small, but not always negligible, disturbance of gravity must be added to the gravitational force. Thus, the total force of the tide at a given point can be expressed by

$$F = \frac{3kM\rho}{r^3}\left[\sin^2\theta\cos^2\theta + (\cos^2\theta - 1/3)^2\right]^{1/2} = \frac{kM\rho}{r^3}\sqrt{3\cos^2\theta + 1}.$$

From this formula it is evident that the tide force is proportional to the mass of the planet M and inversely proportional to the cube of the distance r. For this reason only the sun and the moon have an effective influence on the tide motion. Furthermore, the tide's force depends on the radius of the earth ρ. If P is located on the line connecting the center of the earth with that of the planet, the previous formula simplifies to

$$F = \frac{2kM\rho}{r^3}.$$

The tide's force has an associated potential:

$$V = \frac{kM\rho^2}{2r^3}(3\cos^2\theta - 1)$$

depending on the angle θ of the vertical component of the sea tide. In the static theory, the elevation η of the sea surface is connected to the tide potential by

$$\eta = \frac{V}{g},$$

where g is the gravity acceleration. The change in the sea level at a point on the earth's surface can be expanded in a series of sinusoidal periodic components. These are originated by the tide force due to the planets, during their absolute and relative motion, at the considered point. The most significant frequencies, connected with the relative and absolute motion of the planets, present in the tide phenomena are:

- Absolute motions
  1. Earth rotation around its own axis in relation to the fixed stars (sidereal day) with a period of 23.93 hours.
  2. Revolution of the moon around the earth (tropical month or period of moon declination) with a period of 27.32 days.
  3. Revolution of the earth around the sun (solar year) period of 365.24 days.
  4. Revolution of the axis of the moon's orbit around the earth with a period of 8.85 years.
  5. Cycle of the moon declinations (oscillation of the plane of the moon orbit with respect to the plane of the earth orbit) with a period of 18.6 years.
- Relative motions
  1. Movement of the sun in relation to a point on the earth with a period of 24 hours (average solar day).
  2. Movement of the moon in relation to a point of the earth with a period of 24.8 hours (moon day).
  3. Movement of the moon in its orbit in relation to the motion of the orbit's axis with a period of 27.55 days (anomalistic day).
  4. Movement of the moon with respect to the sun seen from the earth with a period of 29.53 days (synodic month).

Besides the frequency of the various harmonic components of the tide's force, the theory is also able to forecast the amplitude and phases of the various components. But these two quantities deviate more or less evidently from the theoretical values and each tidal component has the form

$$\eta_i = a_i \cos(\omega_i t + \rho_i).$$

While the value of ω_i (different for each component) is determined by the theory, the values of a_i and ρ_i, also called harmonic constants, must be obtained from the observations by means of mathematical filtering of the series of observed values of the sea level at the tide-gauge stations. The longer the available observation series, the larger the number of detectable components (harmonic constants). Once the harmonic constants have been found, it is possible to construct for any future time the values of each component, and their sum gives the effective value of the tide at the required time. In some conditions (low sea level, matting, bays, etc.) the simple theory explained above is no longer sufficient. It is also necessary to take into account the harmonics of order higher than the daily ones, due to the characteristic oscillations of the examined basins or to particular hydrodynamic phenomena. The tides in the open sea do not have relevant amplitudes and are limited to a few decimeters. The amplitude is enhanced for a resonance effect in an area of shallow seas or near coast basins with a particular morphology. By means of mathematical models it is now possible, given the basin's true shape, to solve the dynamic problem of the tides by giving amplitudes and real phases of the components. Static theory can establish only the frequencies. In the semiclosed seas the tides should be of small amplitude because the horizontal component of the tide force does not succeed in generating horizontal motions which are the ones that, on large scales, create observable vertical motions. It might happen that the tide, acting near the open entrance of a basin, determines, as a resonance effect, tide oscillations which often have the same characteristics as normal tides but with amplitudes and phases very different from those derived in the static theory. The Coriolis deviation acts on the tides' currents which are generally weak to the order of a few centimeters per second, but can get stronger in some confined waters, in such a way that a transversal tide component, with respect to the original motion, develops. Rotation movements arise so that, both in ocean basins and in the restricted areas, it is possible to have, in different positions in the daily and semidaily components, convergent flow lines along which the tides have the same phase (the lines with maximum tides at the same time are called cotidal lines). In general, the tides have stationary characteristics and the cotidal lines converge at the points that are called amphidromic, where the tide is constantly zero. The distribution of such nodes depends on the shape of the basin as well as on the Coriolis force. The free oscillations of semiclosed basins are called seiche and they overlap with the tide oscillations, changing their total amplitude. Seiches have different periods, each case depending essentially on the basin geometry. The cause of the occurrence of such effects is to be found mainly in the wind action and in the atmospherical depression passing through the basin. The wind pushes the water and if its movement is toward the coast, this creates an increase in the sea level. As the wind drops, the level tends to the equilibrium through oscillations decreasing over time. Analogously, the whirlpool provoked by the transit of an atmospheric depression causes free and persistent oscillations in the basin. Significant changes in sea level in small semi-closed basins might also be caused as a resonance effect by the oscillations induced by external forces arising in larger connected basins (forced oscillations) or by the action of strong and persistent winds which build up water on the coast.

2.5 Tide Gauges

The most common type of tide gauge used worldwide is the stilling well with float. Other technologies, such as those measuring pressure fluctuations, are also com-

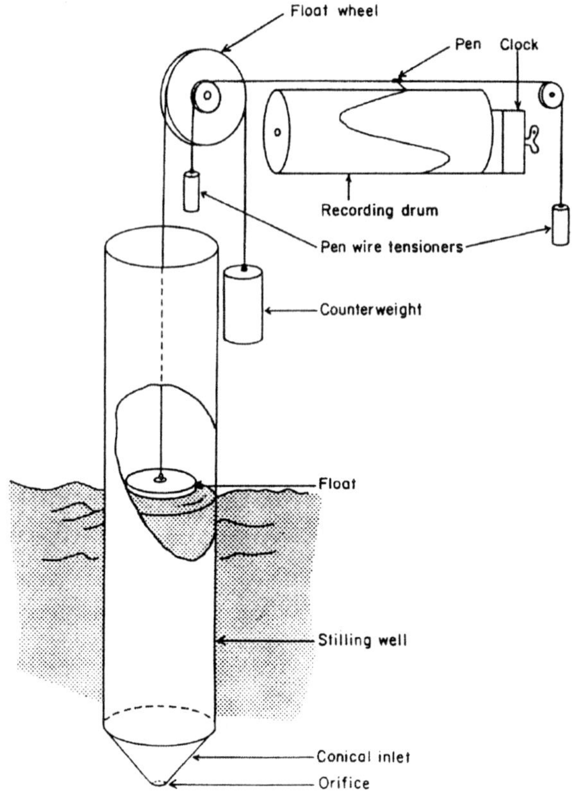

Fig. 2.2. Tide gauge with float.

monly used. Acoustic techniques, measuring the covering time of the impulse reflected from the air-sea interface with an automatic correction for the change in sound velocity, have also improved. In addition, a great number of technical developments have been tried for the sensors and new mechanical and electronic techniques have improved their resolution, reliability and robustness. Microwave and laser techniques have been employed for wave measurement but the sensors are not yet sufficiently accurate for measuring the sea level. In each kind of system the long-term stability in the measurement of the sea level can be optimized by an efficient use of data control.

Tide gauges with float

The simplest form of installation of a tide gauge with float is like the scheme shown in Fig. 2.2. The float on the water surface is connected to a counterbalance by a cable passing through a pulley. As the float rises with the water, the weight goes down accordingly, and the thread makes the pulley rotate at an angle proportional to the change in water level. The second pulley connected to the same rod will rotate at the same angle as the previous one but, since its diameter is smaller, the pen attached to the chord sliding on the pulley will make a shift smaller than the one of the float, even though of the same type. The pen is equipped for writing on a graph in order to describe the movement of the float and by this means, its movement at the surface, though on a reduced scale. The reduction ratio is given by the relationship between the movements of the pen and the water. In this example it is determined by the ratio between the diameters of the two pulleys. If the paper moves with respect to the trajectory of the pen with a fixed velocity, a continuous graph of the sea level as a function of time is obtained. The float operates inside a well which prevents the float sensing the wind and minimizes oscillations due to the short-period waves.

Ultrasound tide gauges

The ultrasound tide gauges measure the distance between the free surface and the transducer by establishing the time employed by the impulse vertically reflected by the sea-air interface before returning to the transducer itself. Such instruments operate at a frequency of about 50 kHz with a cone of 4–6° width. This kind of measurement can be made outdoors with the acoustic transducer mounted vertically but the signal might be disturbed in certain situations. In order to have a continuous operation, the acoustic impulses are usually contained inside a vertical tube or a well equipped with a graduated scale.

Pressure tide gauges

The principle of all measuring systems of this type is to convert the level of the hydrostatic pressure of a water column recorded at a free surface at a certain point. Bubble pneumatic systems deserve a particular mention. Their design is as follows: air is sucked in through a narrow tube fixed over the tide level in such a way that the flow is low and the air pressure equals the hydrostatic pressure with the atmospheric pressure. An instrument for registering the pressure, connected to the injection tube, can register the changes in water level as changes of pressure.

Reference level

The reference level of a tide gauge is the plane of zero measurement of the instrument. This reference plane can be taken as:

- The minimum level of the low syzygy tides. Below this level the depth of the nautical maps is measured while above it the tide levels are measured: it represents a horizontal plane over a limited area and its height will change near the coastal line according to the maximal tide change in the considered places.
- An imaginary reference level (ordinance datum) extending over a large area, in the Italian case over the entire national territory.

The raising of the plane is obtained from the observations of the sea level at a reference point over a long period (the fundamental tide gauge of Genoa); the official reference generally almost coincides with the mean level of the sea. The measure of this level must be referred to a datum point in the neighborhood which is called the datum point of the tide gauge. The datum point must be connected to the national leveling network so that the measures can be referred to a national leveling reference. The measures must be repeated indicatively every year in order to take into account the height changes in the network of the reference datum points. Moreover, the connection with the national network supplies only a nominal height for the tide gauge and is not a determining factor in establishing the movements of the earth's crust at the site. Normally these movements are of the order of a few millimeters per year. Recently, improved modern techniques related to the geodesy of the space and of the absolute gravity have supplied new methods for fixing the tide gauge datum points.

2.6 The National Sea Level Measurement Network (RMN)

The National Sea Level Measurement Network (RMN) consists of 26 new stations uniformly distributed along the Italian coasts (see Fig. 2.3) and located mainly in the following port complexes: Trieste, Venezia Lido, Ancona, Ravenna, Ortona, Vieste, Bari, Otranto, Taranto, Crotone, Reggio Calabria, Messina, Catania, Porto Empedocle, Lampedusa, Palermo, Palinuro, Salerno, Napoli, Cagliari, Carlo Forte, Porto Torres, Civitavecchia, Livorno, Genoa and Imperia.

In these stations mechanical tide gauges with paper registration functioned for many years under the management of the Maritime Corp of Engineers. In 1986 new electronic tide gauges with float and local registration on EPROM chips were installed. The new RMN stations are based on two measuring instruments of the tide: the main one is an ultrasound and a tubular guide for the wave packet with temperature compensation; the secondary one contains float and paper registration for the punctual validation of the measures, for the analysis of particular events or phenomena and for data recovery in case the main instrument fails. The tide gauges are referred to a quoted datum point with a high precision leveling, suitably verified on the basis of the nearest National official datum point. The stations are also equipped with an anemometric sensor (velocity and wind direction at 10 meters above the ground), barometric sensor and sensors for the air temperature and the water temperature. All the stations are equipped with a local system of data management and storage and with a device of real-time transmission to the headquarters of the Marine Service of APAT in Rome.

2.7 Conclusions

Waves and tides are the most perceptible physical phenomena of sea dynamics, both have a great influence on human activities. The coastal equilibrium, port construction

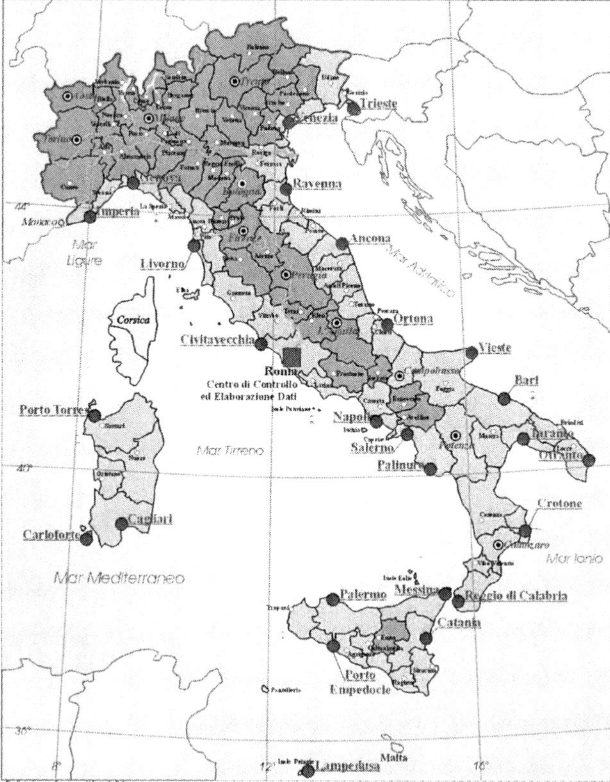

Fig. 2.3. RMN.

and shipping, so important for commerce and exchange, strongly depend on understanding such phenomena. Their phenomenology is explained using the governing physical laws and the local characteristics due to related morphological, meteorological or hydrodynamical details. The basic information comes from long-term measures made according to international standards. Using this data it is possible to apply theoretical laws to the local characteristics. For a good description of the long-term aspects of the phenomena it is necessary to take measurements over a long period of time, even decades, and to obtain time series sufficiently continuous to apply statistical estimates. This is particularly true for the extreme-event analysis of waves and sea levels, when the lack of records for extreme events like a strong sea storm deprives the statistical analysis of the distribution of the maxima of important values and seriously alters the results. Another situation in which such a lack of records creates problems is the case of forecasting based on postprocessing techniques. The measured data are introduced in real or almost real time in the algorithms in order to correct possible errors due to mathematical formalism. In this case the continuity of measurements is also essential and the missing measurements must be replaced by values reconstructed using statistical techniques. Artificial neural networks (ANNs) provide a useful tool for this reconstruction. The algorithm is prepared by means of a learning process, explained in Chapter 4, and a testing process. Once the procedure has given reliable results, the ANN method can be applied for reconstructing the missing values with a remarkable level of reliability. Using ANNs it is possible to also use as input the values of the time series of stations other than the considered one. The only drawback is that the measurements from these stations must be correlated with the time series of the analyzed stations because both are measuring the same physical phenomena. The rest of the book is devoted to explaining the theoretical basis and the practical implementation of the method used to solve the problems of missing gaps for wave and tide time series. In Chapter 7 the reader can find a detailed description of this work.

3

The Wave Amplitude Model

In this chapter we explain the main ideas and open problems of the model of wind waves. It is important to include this presentation in our book since we will compare the results of neural network (NN) reconstruction with those of the wave amplitude model (WAM) model. This comparison is done to check the order of magnitude of the significant wave height (SWH) reconstructed by means of the NN. Moreover, an understanding of this chapter is useful to obtain a good comprehension of the sea phenomena such as the sea evolution, the interaction among the sea and the wind and the formulation of the problem of dissipation forces.

3.1 The WAM

In this section we expose the main ideas underlying the WAM. Since we compare our results on time-series reconstruction made using the NN with the reconstruction made with this model, we also review the basic facts which lead to the WAM approach. We also perform a critical review of these ideas. The WAM (Komen [32], Janssen [31]) is the wave model that is used for ocean wave forecasting at the ECMWF (European Center for Medium Range Weather Forecasts). This is the first model that solves the complete action density equation, including non-linear wave-wave interactions. It describes the rate of change of the spectrum due to advection, wind input and dissipation due to non-linear interactions. The wave spectrum gives the distribution of wave energy over frequency and direction and gives a complete specification of the sea state. At present two versions of the WAM are used operationally in global and regional applications at ECMWF: the global model and the limited area model. The global model has an irregular latitude grid with a resolution of 55 kilometers, an advection time step of 20 minutes (the same as for the source term integration) and a wave spectrum of 25 frequency bins and 12 directions. The limited area model has a resolution of 28 kilometers, shallow water effects are included and the advection and source time step is 10 minutes. The wave spectrum is approximated with discrete variables of 25 frequency bins and 24 directions. The limited area model covers the North Atlantic, Norwegian Sea, North Sea, Baltic Sea, the Mediterranean and the Black Sea. Coupled to the ensemble prediction system, fifty wave forecasts are generated from fifty ensemble forecasts. From this a wide range of information can be derived, for example, the probability of significant wave heights higher than 4 meters. A large number of variables take part in the wave growth as will be clarified in the presentation. The frequency ω, the wind speed at 10 m height U_{10} or the friction velocity u_*, the gravity acceleration g, the viscosity, the surface tension, the air and water density and the Coriolis parameter f are only some of the variables which are considered in the idealized situation of duration-limited waves. In the following we treat the case of duration-limited waves when a uniform and steady wind has been blowing over an unlimited ocean for a time t after a sudden onset. Conditions of duration-limited growth are difficult to fulfill in practice. Two other idealized cases are important: the case of fully developed waves and the fetch-limited case. The case of fully developed waves occurs when a uniform and steady wind has blown over an unlimited ocean long enough for the wave field to become independent of time. We have the fetch-limited case when a uniform steady wind has been blowing from a straight shoreline long enough for the wave field at a distance (fetch) from the upwind shore to become independent of time. In the following sections we are going to explain briefly the assumptions, the main equations and the variables of the model starting from the very beginning of the Euler equations for a two-dimensional ocean free air-water interface.

3.2 The Free Surface Problem

We consider the sea and the air in the two-dimensional vertical plane $\mathbf{x} = (x, z)$. We assume that the horizontal scale of the sea is much larger than its depth (the shallow water approximation) and the height of the surface of the sea (SWH) is described by a function $\eta(\mathbf{x}, t)$, where $t \geq 0$ is the time variable. The water is considered to be incompressible. Let $\mathbf{u}$ be the velocity of the fluid at the point $\mathbf{x}$. Then the general equations governing the motion of this system are

$$\nabla \cdot \mathbf{u} = 0 \tag{3.1}$$

$$\left(\frac{\partial}{\partial t} + \mathbf{u} \cdot \nabla\right)\mathbf{u} = -\frac{1}{\rho}\nabla P - \mathbf{g} + \nabla \cdot \tau, \tag{3.2}$$

where $\mathbf{g}$ is the gravity acceleration, P is the pressure and ρ is the density. $\rho = \rho_a$ in the air, $z > \eta(\mathbf{x}, t)$, $\rho = \rho_w$ in the water, $z < \eta(\mathbf{x}, t)$, where ρ_a, ρ_w are, respectively, the density of the air and of the sea. τ is the tangential stress generated by the wind on the sea surface. We have to add to these equations the kinematic boundary conditions connecting the total derivative of $\eta(\mathbf{x}, t)$ with respect to the time t with the vertical velocity w of the fluid at the sea surface,

$$\left(\frac{\partial}{\partial t} + \mathbf{u} \cdot \nabla\right)\eta = w.$$

The velocities and forces are continuous at the sea-air interface. The boundary conditions on the solution $\mathbf{u}$ of the system (3.1, 3.2) must be introduced. At the bottom of the ocean the normal component of the velocity w must vanish. In practice, for the ocean and wave dynamics it is better to consider the case of Euler equations in which viscosity and stresses are neglected. When considering the non-linear phenomena of wave growth we must take into account a stress induced from the waves themselves. In the approximation of Euler equations the continuity of the stress at the interface is no longer required. Another assumption that we make is that $\rho_a/\rho_w \ll 1$ and so the air motion can be disregarded in a first approximation. But in the derivation of the final equations the effect of the ratio ρ_a/ρ_w will be taken into account. The third important assumption is that the velocity vector is irrotational. In the framework of Euler equations the vorticity remains zero if it is zero initially. In the following we use the symbol ∇ for the two-dimensional gradient $(\partial/\partial x, \partial/\partial y)$, introducing the coordinate y in the analysis since we really need to describe a three-dimensional evolution. We use the symbol $\triangle$ for the two-dimensional Laplacian $(\partial^2/\partial x^2)+(\partial^2/\partial y^2)$. As a consequence of these assumptions we can introduce a velocity potential Ψ:

$$\mathbf{u} = \nabla\Psi. \tag{3.3}$$

Then the equations of motion become

$$\triangle\Psi + \frac{\partial^2}{\partial z^2}\Psi = 0 \tag{3.4}$$

$$\frac{\partial\eta}{\partial t} + \nabla\Psi\cdot\nabla\eta = \frac{\partial}{\partial z}\eta, \quad z = \eta \tag{3.5}$$

$$\frac{\partial\Psi}{\partial t} + \frac{1}{2}(\nabla\Psi)^2 + \frac{1}{2}\left(\frac{\partial\Psi}{\partial z}\right)^2 + g\eta = 0, \quad z \le \eta. \tag{3.6}$$

(3.6) is the Bernoulli equation which we will use only on the surface of the sea. The boundary condition at the flat bottom of the ocean $z = -D$ expresses the constraint that the fluid does not pass through it:

$$z = -D, \quad \frac{\partial}{\partial z}\Psi = 0.$$

Another useful way to formulate the water wave problem is the variational principle of Luke [39] which allows us to derive the equation of motion from a Lagrangian formulation:

$$\delta\int \mathcal{L}\,dxdydt = 0 \tag{3.7}$$

with

$$\mathcal{L} = -\rho\int_{-D}^{\eta} dz\left[\frac{\partial}{\partial t}\Psi + \frac{1}{2}(\nabla\Psi)^2 + \left(\frac{\partial\Psi}{\partial z}\right)^2 + gz\right]. \tag{3.8}$$

From the variational principle (3.7) it is possible to get the system (3.4), (3.5), (3.6) including the boundary conditions. The variational formulation also gives, as usual for the Lagrangian formulations, the conservation of an energy functional $\mathcal{E}$:

$$\mathcal{E} = \frac{1}{2} g\rho \int d\mathbf{x}\eta^2 + \frac{1}{2}\rho \int d\mathbf{x} \int_{-D}^{\eta} dz \left[(\nabla\Psi)^2 + \left(\frac{\partial\Psi}{\partial z}\right)^2 \right]. \tag{3.9}$$

The equations of the WAM describe the motion of a local group of waves in the ocean or sea. The statistical information is also an important element of this model and the main quantity of the model equations, the "action density," is connected with the spectral properties of the waves. We will first indicate the procedure which permits us to arrive at the model starting from the dispersion relations for one wave only in the linearized theory; then we will consider the packet of multiple waves.

3.3 The Linearized System

If we consider only fluctuations of η small with respect to the depth of the basin D we can neglect all the second-order terms in the previous system and consider the linearized system:

$$\Delta\Psi + \frac{\partial^2}{\partial z^2}\Psi = 0 \tag{3.10}$$

$$\frac{\partial\eta}{\partial t} + \nabla\Psi \cdot \nabla\eta = \frac{\partial}{\partial z}\eta, \quad z = \eta \tag{3.11}$$

$$\frac{\partial\Psi}{\partial t} + \frac{1}{2}(\nabla\Psi)^2 + \frac{1}{2}\left(\frac{\partial\Psi}{\partial z}\right)^2 + g\eta = 0, \quad z \leq \eta. \tag{3.12}$$

We impose the boundary between air and water to be the $z = 0$ hyperplane and take the same boundary conditions at the sea bottom, i.e.,

$$z = -D, \quad \frac{\partial}{\partial z}\Psi = 0$$

and we get the following system:

$$\Delta\Psi + \frac{\partial^2}{\partial z^2}\Psi = 0 \tag{3.13}$$

$$\frac{\partial\eta}{\partial t} = \frac{\partial\Psi}{\partial z}, \quad z = 0 \tag{3.14}$$

$$\frac{\partial\Psi}{\partial t} + g\eta = 0, \quad z = 0 \tag{3.15}$$

$$\frac{\partial}{\partial z}\Psi = 0, \quad z = -D. \tag{3.16}$$

In this case we can look for a simple plane wave solution (gravity wave)

$$\theta = \mathbf{k} \cdot \mathbf{x} - \omega t$$
$$\eta = a \exp(i\theta)$$
$$\Psi = aZ(z) \exp(i\theta), \tag{3.17}$$

i.e., we consider a plane wave describing the evolution of the sea elevation (SWH) with phase θ, wave vector $\mathbf{k}$, angular frequency ω and velocity potential Ψ depending on the quota z.

Substituting in the system one gets

$$Z'' - k^2 Z = 0 \tag{3.18}$$
$$k = |\mathbf{k}| = \sqrt{k_x^2 + k_y^2}$$
$$Z' = 0 \quad z = -D.$$

From these equations and conditions we easily have

$$\Psi = -ia\frac{g}{\omega}\frac{\cosh k(z + D)}{\cosh(kD)}$$
$$\omega^2 = gk \tanh(kD). \tag{3.19}$$

The latter equation is obtained by substituting the expressions for η and Ψ in the equations (3.15), (3.16). The relation between ω and k is the dispersion relation. It is characteristic of the waves (called gravity waves) obtained in the linear approximation. For small k it gives the well-known relation $v \sim \sqrt{gD}$ where $v = \omega/k$ is the velocity of propagation of the wave. It can also be used to estimate the propagation velocity of a tsunami since the linear theory is considered to be valid for tsunami propagation in deep waters. For a more realistic description of waves propagating in the ocean it is necessary to consider a group of waves.

3.4 Wave Packet

The main problem with the linearized approach of the last section is that it considers only one wave of small amplitude. A realistic model has to consider the situation of packets of waves of arbitrary amplitude propagating together. So we shall consider a wave description as before but with the coefficients being a slowly varying function of time and position. Thus, the starting formula for the wave altitude η is

$$\eta = a(\mathbf{x}, t) \exp i\theta(\mathbf{x}, t) + cc \tag{3.20}$$

where cc means complex conjugate and the phase $\theta(\mathbf{x}, t)$ and the amplitude $a(\mathbf{x}, t)$ are slowly varying functions of space and time. The frequency ω and the wave vector $\mathbf{k}$ now are functions of $\mathbf{x}, t$. This description is necessary if non-uniform sea bottoms are considered. In this case the dispersion relation connecting $\mathbf{k}$ and ω is of the type

$$\frac{\partial \mathbf{k}}{\partial t} + \nabla\omega = 0. \tag{3.21}$$

This equation is a consistency relation expressing the conservation of the number of crests of the waves. It is possible to apply a classical averaging procedure introduced by Whitham ([70]). We introduce this formula for the wave amplitude η into the Lagrangian density (3.8) and average the Lagrangian over the slowly varying phase θ:

$$\langle \mathcal{L} \rangle \equiv \frac{1}{2\pi} \int_0^{2\pi} d\theta \mathcal{L}. \tag{3.22}$$

Neglecting the wave-induced currents, we obtain an average Lagrangian $\langle \mathcal{L} \rangle$ which depends on $\omega, \mathbf{k}$ and a:

$$\langle \mathcal{L} \rangle = \Upsilon(\omega, \mathbf{k}, a), \tag{3.23}$$

where a is the wave amplitude entering (3.20), $\omega = -\partial\theta/\partial t$ and $\mathbf{k} = \nabla\theta$. For simplicity, we use the same symbol $\mathcal{L}$ for the averaged Lagrangian. Hence, the appropriate evolution equations follow from the variational principle

$$\delta \int dx dt \mathcal{L}(\omega, \mathbf{k}, a) = 0, \tag{3.24}$$

where the dependence of $\mathcal{L}$ on the variables x, t is through the functions $\omega, \mathbf{k}, a$. The variation with respect to the amplitude a gives the dispersion relation:

$$\frac{\partial}{\partial a} \mathcal{L}(\omega, \mathbf{k}, a) = 0. \tag{3.25}$$

The variation with respect to the phase θ gives the evolution equation for the amplitude

$$\frac{\partial}{\partial t} \mathcal{L}_\omega - \nabla \mathcal{L}_{\mathbf{k}} = 0, \tag{3.26}$$

where $\mathcal{L}_\omega, \mathcal{L}_{\mathbf{k}}$ are the partial derivatives of $\mathcal{L}$ with respect to ω and the components of the vector $\mathbf{k}$. To these equations one must add the dispersion relation. This set of equations describes the evolution of the slowly varying wave group. If we introduce the transport velocity

$$\mathbf{u} = -\frac{\mathcal{L}_k}{\mathcal{L}_\omega}$$

the equation (3.26) becomes

$$\frac{\partial}{\partial t} \mathcal{L}_\omega + \nabla \cdot (\mathbf{u} \mathcal{L}_\omega) = 0. \tag{3.27}$$

This equation defines the evolution of the action density $\mathcal{L}_\omega$. Using these equations and expanding the averaged Lagrangian (3.23) in the parameter $\epsilon = 2\rho g |a|^2$ we can rewrite it:

$$\mathcal{L} = \frac{1}{2} \epsilon R(\omega, \mathbf{k}), \tag{3.28}$$

where

$$R = \frac{(\omega - \mathbf{kU_0})^2}{2gkT} - 1, \tag{3.29}$$

and $T = \tanh(kD)$ and U_0 is a reference vector of velocity. So if we define $\sigma = \sqrt{gkT}$, by the dispersion relation (3.25) we obtain $R = 0$. From this,

$$\omega = \mathbf{kU_0} \pm \sigma, \tag{3.30}$$

where the velocity of the current $\mathbf{U_0}$ and the depth D are allowed to be slowly varying functions of space and time. Finally, we can define the action density for the case of a wave packet:

$$N = \mathcal{L}_\omega = \frac{\epsilon}{\sigma} \tag{3.31}$$

and the evolution equation becomes

$$\frac{\partial}{\partial t} N + \nabla \cdot (\mathbf{v_g} N) = 0, \tag{3.32}$$

where $\mathbf{v_g}$ is the group velocity $\partial\omega/\partial\mathbf{k}$. This equation is called the *action balance equation*. It is the most important equation of the model since all the applications start from it to arrive at physical results. However, it is necessary to also introduce the random nature of the process of evolution of the packet of ocean waves. This is done in the next section where the final model equations are given. We note the structure of transport phenomena of equation (3.32) from which the equations of WAM are derived. There are still terms to be added on the right-hand side and they are discussed in the next section. The situation of equations (3.32) and (3.33) is similar to that of the Boltzmann equation since there is no rigorous derivation in their case as well and the terms to be added are very similar to the interaction integral appearing in the Boltzmann equation. The difference being, so far, the interpretation of the function N (the Boltzmann equation describes the evolution of gases and fluids and it is also a transport equation).

3.5 The Action Balance Equation

Equation (3.32) has been derived neglecting the non-linear interaction among the packets of waves, the force of the wind and the terms connected to the loss of energy. The full equation is

$$\frac{\partial}{\partial t} N + \nabla \cdot (\mathbf{v_g} N) = \left.\frac{\partial N}{\partial t}\right|_w + \left.\frac{\partial N}{\partial t}\right|_n + \left.\frac{\partial N}{\partial t}\right|_d, \tag{3.33}$$

where the first term on the right-hand side of the equation is the action of the wind, the second is the non-linear interaction and the third is the dissipation term. Before briefly discussing these terms we have to introduce the statistical meaning of the surface elevation $\eta(\mathbf{x}, t)$. It is natural to assume that the surface elevation is a random variable and, more exactly, a random process due to the randomness of the wave motion. More precisely, $\eta(\mathbf{x}, t)$ is supposed to be a translation invariant Gaussian random field at fixed time t. This means that the probability distribution of the variables $\eta(\mathbf{x}_1, t), \ldots, \eta(\mathbf{x}_n, t)$ for arbitrary points $\mathbf{x}_1, \ldots, \mathbf{x}_n$ is completely determined by the knowledge of the first two moments:

$$\langle \eta(\mathbf{x}, t) \rangle, \quad \langle \eta(\mathbf{x}_1, t)\eta(\mathbf{x}_2, t) \rangle$$

and also that the correlation $\langle \eta(\mathbf{x}_1, t)\eta(\mathbf{x}_2, t) \rangle$ is a function only of the distance $\mathbf{x}_1 - \mathbf{x}_2$ (the latter because of translation invariance). Here the symbol $\langle \cdot \rangle$ means average with respect to the probability distribution of this Gaussian field. Furthermore, if dissipative effects are strong enough, the waves become independent very rapidly and so this correlation decays exponentially. But this statement can be accepted only if we consider the height of the sea $\eta(\mathbf{x}, t)$ as the sum of a deterministic component (some average of the field or some first deterministic Fourier component) plus a fluctuation which is a Gaussian random field of small amplitude; otherwise, it is in contradiction with the fact that $\eta(\mathbf{x}, t)$ is a solution of the evolution equations introduced in the first section. A derivation of the central limit theorem (CLT) for the wave height can be made ([60]) under the hypothesis of this decomposition and supposing the random components to be of the form of the sum of CLT. From this argument it follows that the correlation function $C(\xi)$ of the significant wave height (SWH) is the fundamental object

$$C(\xi) = \langle \eta(\mathbf{x} + \xi, t)\eta(\mathbf{x}, t) \rangle.$$

It is natural to introduce the Fourier transform of the correlation function:

$$F(\mathbf{k}) = \frac{1}{(2\pi)^2} \int d\xi \exp[i(\xi \cdot \mathbf{k})] C(\xi).$$

This representation generalizes easily to the more interesting case of a wave packet:

$$\begin{aligned} \eta(\mathbf{x}, t) &= \int_{-\infty}^{+\infty} \hat{\eta}(\mathbf{k}) \exp[i(\mathbf{k} \cdot \mathbf{x} - \omega t)] + cc \\ \langle \hat{\eta}(\mathbf{k})\hat{\eta}(\mathbf{k}') \rangle &= 0 \\ \langle \hat{\eta}(\mathbf{k})\hat{\eta}^*(\mathbf{k}') \rangle &= |\hat{\eta}(\mathbf{k})|^2 \delta(\mathbf{k} - \mathbf{k}'), \end{aligned} \tag{3.34}$$

where we have used translation invariance of the process. It follows easily then that the integral of the Fourier transform of the correlation function is equal to the variance of the SWH:

$$\langle \eta^2 \rangle = C(0) = \int d\mathbf{k} F(\mathbf{k}). \tag{3.35}$$

At this point an equation connecting the action density with the spectral function of the process $\eta(\mathbf{x}, t)$ is introduced in the classical texts about WAM:

$$N(\mathbf{k}) = \frac{F(\mathbf{k})}{\sigma},$$

where $\sigma = \sqrt{gk \tanh(kD)}$. But with this identification the action balance equation becomes a transport equation for the flow of the spectral function. The connection with the moving boundary problem associated with the fundamental set of partial differential equations is lost. Thus, it is necessary to improve this approach. We made a rigorous derivation of an equation for the SWH η, using the approach of pseudo differential operators, starting again from the free boundary problem of the first section but there is no stochastic behavior in it (Dobrokhotov, Tirozzi and Tudorovsky [61]). Instead the approach followed in this book is to make the statistical analysis of the waves using NNs. This approach has been shown to be useful as can be seen from the results of Chapter 7, where a comparison of the output of NN and WAM is made. This success suggests that we also consider NN as part of the tools to use for a reformulation of the problem of the description of SWH. We conclude this section by discussing the three contributions to the action density equation.

1. *Action of the wind.*

The rate of change of the action density due to the wind is

$$\left.\frac{\partial N}{\partial t}\right|_{w} = 2\gamma N, \tag{3.36}$$

where γ is the growth rate of the waves

$$\gamma = \mathrm{Im}(\omega).$$

The growth rate γ is proportional to the second derivative with respect to z of the horizontal component of the wind multiplied by the frequency of the waves and by the ratio of the density of the air with respect to the density of the water. We introduce the following definitions:

$$\begin{aligned}
\mathbf{U}_0 &= U_0(z\mathbf{e}_x)\\
c &= \frac{\omega}{k}\\
W &= U_0 - c\\
\epsilon &= \frac{\rho_a}{\rho_w}\\
\chi &= \frac{w}{w(0)}.
\end{aligned} \tag{3.37}$$

$\mathbf{U}_0$ is the horizontal velocity of the air, c the phase velocity of the wave with wave vector k, W is a kind of relative velocity, ϵ is a perturbation parameter equal to the ratio of the density of air with respect to the density of the water and χ is the ratio between the vertical velocity at the generic point z divided by the vertical velocity at the point $z = 0$, $w(0) \neq 0$. We indicate with the subscript c all the quantities evaluated at the critical height z_c ($W_c = 0$). Then the coefficient of the wind forcing γ (3.36) is given by the formula

$$\frac{\gamma}{\epsilon\omega} = -\frac{\pi}{2k}\frac{W_c''}{|W_c'|}|\chi_c|^2. \tag{3.38}$$

It follows from this formula that the gravity waves generated by a wind with a negative profile of the curvature W_c'' are unstable. This result has been shown by Miles ([46], [47]).

2. *Non-linear interaction.*

The non-linear terms of the four-wave interaction can be described in a way similar to the scattering of waves and particles, and the relative contribution is the same as the collision term in the Boltzmann equation. If we consider four waves with frequencies $\omega_1, \ldots, \omega_4$ and wave vectors $\mathbf{k}_1, \ldots, \mathbf{k}_4$ there is a conservation of momenta and energies as in particle collisions:

$$\begin{aligned} \omega_1 + \omega_2 &= \omega_3 + \omega_4 \\ \mathbf{k}_1 + \mathbf{k}_2 &= \mathbf{k}_3 + \mathbf{k}_4. \end{aligned} \tag{3.39}$$

An analogous system of equations for the scattering of three gravity waves does not hold because it is inconsistent with the dispersion relation for gravity waves $\omega = \sqrt{gk}$. The conservation of the sum of momenta $\mathbf{k}_i$ is not compatible with the conservation of the frequencies ω_i due to the structure of the energy-impulse relation. The situation is even more difficult in the shallow water case with a non-uniform bottom where both the impulse and energy of the waves depend on the position. We call the action density of the four waves with analogous indices $N_1, \ldots, N_4$; then the term coming from the contribution of non-linear interaction among the waves is

$$\begin{aligned} \left.\frac{\partial N}{\partial t}\right|_n = 4 \int & d\mathbf{k}_2 d\mathbf{k}_3 d\mathbf{k}_4 T^2_{1,2,3,4}\delta(\omega_2 + \omega_3 - \omega_4 - \omega_1) \\ & \times \{N_2N_3(N_1 + N_4) - N_1N_4(N_2 + N_3)\}, \end{aligned} \tag{3.40}$$

where T depends on the wave parameters $\mathbf{k}_i$ and ω_i is the analog of the scattering amplitude or cross section for the four-wave interaction. Clearly, the integral on the right-hand side is the analog of the collision integral for the

Boltzmann equation. Details of its derivation may be found in Hasselmann ([22]) and Komen ([32]).

3. *Dissipation.*

The least-understood term of equation (3.33) is the dissipation source function. Waves may lose energy continuously by viscous dissipation and by the highly intermittent process of wave breaking. The continuous slow drain of wave energy by viscosity is well understood but is important only for gravity waves with a wavelength of 1 cm and becomes insignificant for longer waves. Understanding the wave breaking process is very important for this theory but up to now there has not been much progress in this direction. Hasselmann ([23]) proposed a form of the dissipation source function. Introducing the mean frequency $\langle\omega\rangle$ and mean wave number $\langle k\rangle$,

$$\langle\omega\rangle = \frac{\int d\mathbf{k}\omega N}{\int d\mathbf{k}N} \tag{3.41}$$

and a similar relation for $\langle k\rangle$, the dissipation source term takes the following form:

$$\left.\frac{\partial N}{\partial t}\right|_d = -\beta\langle\omega\rangle(\langle k\rangle^2 m_0)^2[(1-\delta)x + \delta x^2]N, \tag{3.42}$$

where $x = k/\langle k\rangle$ and $m_0 = \langle\eta^2\rangle$ is the wave variance. The constant β is unknown and must be adjusted for fitting the experiments. The equation of the action density with these three inputs is integrated for predicting the evolution of the group of waves. The equation has a local character because the spectral density is computed by making the Fourier transform in a finite region around the point of interest. We can say that the WAM still requires much work for a clearer setting and still needs to be generalized to arbitrary bottom configurations. Furthermore, the connection of the WAM with the free surface problem is far from being understood.

4

Artificial Neural Networks

In this chapter we explain some fundamental concepts used in neural networks (NNs) with special regard to the ones applied to forecasts and data analysis. The structure of the back-propagation NN is shown in connection with its use for time-series analysis. The concepts of learning and training errors are explained in some detail and the main types of learning algorithms are exhibited. We also give some exact estimates of the probability that the learning error differs from the training error by more than a small quantity and a priori bounds of the learning error using extreme-value theory for the simple perceptron. Since NNs are also analyzed as a universal approximation of functions, we expose some useful properties of NNs treated in this way in Chapter 5. These topics do not cover all the literature dealing with NN application and theory, but they are good examples of the present situation. Heuristic remarks and rigorously proven statements are also distinguished using mathematical language for exactly proven facts and ordinary language for conjectures and hypotheses. This type of presentation suits the general aim of our book: we want to start from the very beginning of the theory to arrive at the very end of some applications of NNs. In the literature, especially in the applications, there is often confusion between proven facts and conjectures.

4.1 Introduction to Neural Networks

In this section we describe the general features of the application of NNs to data analysis. NNs can be thought of as algorithms for studying and modeling any given set of data. They are a non-standard tool of statistical analysis. The big difference between NNs and the classical models like ARIMA (autoregressive integrated moving average), ARCH (autoregressive conditional heteroskedasticity) and GARCH (generalized autoregressive conditional heteroskedasticity) is the lack of a priori hypotheses usually necessary for the applications of these traditional models. In general, it is possible to study and make predictions on any set of data with NNs [26]. The main hypothesis of the neural algorithms is that the (usual) non-linear relationship which generates the data can be "learned," in a sense, which will be made more explicit below, directly from the sequence of data $\{\underline{x}^\mu, y^\mu\}_{\mu=1}^P$ where $\underline{x}^\mu$ is an N-dimensional vector and y^μ is a scalar, N is the value of the embedding dimension and μ is usually a time index. In order to use the NN the estimate of the embedding dimension of the input vector is made at the beginning of the application. In other words, we suppose that there exists an unknown function, usually non-linear, such that

$$y^\mu = f(\underline{x}^\mu).$$

According to the current terminology of neural networks, $\underline{x}^\mu$ is the input vector and y^μ is the output of the network. The knowledge of the function f allows us to make predictions of the series at a subsequent time $\mu + 1$, μ being a time index and y^μ being the value of the time series at the time $\mu + 1$. Let us be more precise. We restrict ourselves to the particular case of predicting the data at the time $\mu + 1$ for the sake of simplicity, although the relationship which can be extrapolated with the NN is at any time $\mu + n$, where n cannot be larger than the embedding dimension N. Let $X(t), t \in \mathbf{N}$ be the original time series. Then N equals the correlation length defined as the minimum value of k such that $E(X(t)X(t-k)) = 0$ where E is the expectation of the process and we take the case of zero average, i.e., $E(X(t)) = 0\, \forall\, t$, for simplicity. Supposing that it is possible to check that the time series is stationary, at least at the level of the correlation functions, we find that in practice the above correlation $E(X(t)X(t-k))$ is only a function of k. We will see later that there is a value of k such that $E(X(t)X(t-l)) = 0$ for $l \geq k$. This condition is verified by determining the value of k whereby the correlation starts to oscillate around zero with fluctuations less than its variance. This choice of N must be considered as an order of magnitude since sometimes the optimal value of the embedding dimension does not coincide exactly with k. We also have to compare the preceding estimate with the estimate obtained with other methods such as the method of false neighbors or the relative entropy method. We will discuss these approaches in the next section where we also deal with the question of the stationarity in the weak sense, i.e., in the sense of correlation functions. Once we have found the optimal value for N we obtain the following structure of the input vector: $\underline{x}^\mu = (X(\mu), \ldots, X(\mu - N)), \mu \geq N$, since μ is a temporal index, and the output of the network y^μ is the value of the time series $X(t)$ at the time $\mu + 1$. The input-output relationship defined by the network gives an approximation of the value at the time $\mu + 1$ of the process on the basis of knowing the data from time $\mu - N$ up to the time μ. It is also possible to change the order of the indexes and to choose as an input vector the data $X(t)$ taken at N subsequent times $\underline{x}^\mu = (X(\mu), \ldots, X(\mu + N))$ and as the output vector the value of the time series at time $\mu - 1$. The aim of the learning algorithms is to find the best approximation of the function f in order to minimize the approximation error. In this book we apply this search to the time series of sea levels (SL) and to the time series of significant wave heights (SWH) and we show how this application solves some important open problems. We also show the main problem solved by these algorithms: filling the large gaps of missing data that are present in the SWH and SL series due to technical reasons. It is important to have the entire time series to correctly evaluate the statistics of rare or extreme events as shown in Chapter 9. Here we start the description of the general structure of the back-propagation neural networks (NNs) also called artificial neural networks (ANNs). We use the same notation as before $(\underline{x}^{\mu}, y^{\mu})$ with $y^{\mu} = X(\mu + 1)$ and $\underline{x}^{\mu} = (X(\mu), \ldots, X(\mu - N))$. Let also $\phi(\underline{x}^{\mu})$ be the transformation of the network on the input data and ξ^{μ} its output:

$$\xi^{\mu} = \phi(\underline{x}^{\mu}). \tag{4.1}$$

Our problem is to find a ϕ such that ξ^{μ} approximates y^{μ} in the best way for all the P data. The set of P data is called the *training set*. The elements of the training set or the data in general are called *patterns*. The training set of P data is divided into two sets containing L and T data, respectively, $P = L + T$. L and T are arbitrarily chosen but are of the same order of magnitude. The first set (with L data) is used for searching the best ϕ (the *learning set*) and the other (with T data) for testing the chosen optimal ϕ (the *testing set*). An important quantity is the *generalization error* (*GE*). The generalization error is defined as the error performed by the network on a new pattern not belonging to the training set. Thus we have an optimization problem in the sense that we need to minimize the *learning error* (*LE*):

$$LE = \frac{1}{L}\sum_{\mu=1}^{L}(\xi^{\mu} - y^{\mu})^2. \tag{4.2}$$

The *testing error* (*TE*) is of the same form but is evaluated on a set of data not involved in the process of minimization:

$$TE = \frac{1}{T}\sum_{\mu=L+1}^{P}(\xi^{\mu} - y^{\mu})^2. \tag{4.3}$$

The *generalization error* is computed on data ξ^{ν} or patterns not belonging to the training set and has the same form:

$$GE = (\xi^{\nu} - y^{\nu})^2. \tag{4.4}$$

We also have to note that the form of the error is not necessarily the one indicated in the above formulas. Usually the module of the difference is used instead of the square:

$$LE = \frac{1}{L}\sum_{\mu=1}^{L}|\xi^{\mu} - y^{\mu}|. \tag{4.5}$$

In some cases it is better to use a *weighted sum* in the expression of the error with some suitably chosen coefficients as we will show in Chapters 7 and 9. It is interesting to compare the error of the estimate of $X(\mu + 1)$ with the estimates achieved by using traditional statistical algorithms such as the ARIMA model. This will be done in Chapter 8. The NN can be graphically represented for a better understanding of the algorithm. In Fig. 4.1 we show the two-layer perceptron as an example. It is

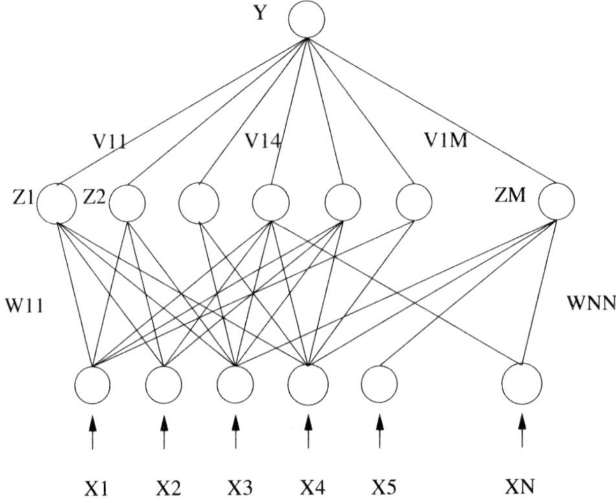

Fig. 4.1. Two-layer perceptron.

possible to also consider multilayer perceptrons which include many intermediate layers, although we will not use them in our analysis.

The terminology associated with this picture is as follows: the first layer is composed of a number of units (neurons) equal to the embedding dimension N. To the jth neuron is associated the jth component of the input vector $\underline{x}^{\mu}$: each neuron is given an input signal equal to a component of the input vector. The first layer is the *input layer* of the NN. Each input neuron is connected with all the neurons of the first layer through a *synaptic interaction* W_{kj}. The number M of neurons belonging to the first layer or to the *hidden layer* is another free parameter which is chosen in order to minimize the learning error LE. The neurons of the first layer send the neurons of the second layer a signal obtained by adding the products of the components of the input vector with the synaptic weights. Thus, the kth neuron receives a signal equal to

$$z_k^{\mu} = \sum_{j=1}^{N} W_{kj} x_j^{\mu}$$

and sends to the upper layer a signal

$$h_k^{\mu} = \sigma(z_k^{\mu} - \theta_k),$$

where $\theta_k \in R$ is a threshold and $\sigma(x)$ is the input-output non-linear function

$$\sigma(x) = \frac{1}{1 + \exp(-\lambda x)}, \qquad \lambda \geq 0, \ x \in R. \tag{4.6}$$

This function describes the non-linear response of the real neurons to the synaptic input. All the concepts used in this situation are suggested by neuroscience ([62],

[30]). The threshold is a quantity which controls the neuronal dynamic: if the sum of the inputs received by a certain neuron is larger than the threshold then this neuron becomes active. However, the analogy is not strict; in fact, in the NN case the output of the kth neuron is a constant varying between zero and one when the signals received from the input layer are larger than the threshold θ_k. However, in the neurobiological case the response of the neuron receiving signals over the threshold is a complicated function of time. The network's output is obtained analogously. If the real-valued variables $V_{11}, \ldots, V_{1M}$ are the synaptic weights between the first layer and the output neuron, then the expression for the function ϕ defined by the NN is

$$\xi^{\mu} = \sigma\left(\sum_{k=1}^{M} V_{1k} h_k^{\mu} - \theta_0\right), \quad \xi^{\mu} = \sigma\left(\sum_{k=1}^{M} V_{1k}\sigma(z_k^{\mu} - \theta_k) - \theta_0\right), \tag{4.7}$$

where $(\theta_0, \ldots, \theta_M) \in R^{M+1}$ is a set of real thresholds. The minimization algorithm of LE acts on the parameters $V_{11}, \ldots, V_{1M}, W_{kj}, k = 1, \ldots, M, j = 1, \ldots, N$ and eventually on the thresholds $\theta_1, \ldots, \theta_M, \theta_0$. It is called the *learning process* in analogy with the learning process of humans and animals. It is known that during an organism's growth the learning of particular responses or actions corresponds to a change of synaptic weights in the neuronal system. In our case the process is much simpler since the learning algorithm is a particular dynamic acting on the space of synaptic weights in order to obtain the best approximation to the function f. The various methods in the literature can be reduced either to the method of steepest descent or to simulated annealing. These algorithms are explained in detail in Section 4.3. Here we offer only some comments on them. The path of steepest descent for the learning error is found by evaluating the derivatives of LE with respect to the variables $V_{11}, \ldots, V_{1M}, W_{kj}, k = 1, \ldots, M, j = 1, \ldots, N$ and eventually with respects to the thresholds $\theta_1, \ldots, \theta_M, \theta_0$. The simulated annealing algorithm is a nonstationary Markov chain acting on the space of the synaptic weights and thresholds which are considered as belonging to some bounded interval and approximated by some discrete set of values. The data are normalized in order to apply the σ function. For our computations we have chosen simulated annealing since this method avoids the phase point being blocked in some local minimum of the function, as described in detail in Section 4.3. There are theorems showing this property; in practice, this algorithm makes it possible to escape from local minima by changing the transition probabilities in a suitable way. Of course, the theorems are not exactly verified in computer simulations since they are never exact. This is due to the finite number of digits used by the computer for representing a real number. Nevertheless, the theoretical arguments and the large number of applications provide good reasons for expecting the required behavior of the method. The possibility of obtaining the absolute minimum is achieved at the expense of a longer time of convergence of the algorithm, although in our case we verified that it is not so long. After the *learning process* is finished, the *testing process* takes place in order to verify that the NN with the synaptic weights found in the first phase is an effective tool for approximating data not considered during the learning process. Thus, we consider the *generalization* property of the NN or analyze its *learning curve*. As we mentioned above, the data set is divided into two disjoint parts. One is the *learning set* made of pairs of data as $(\underline{x}^\mu, y^\mu)$ and the other is used for testing the network on data $\underline{x}^\mu$ not used in the learning procedure. Once the test has been successful, i.e., LE and TE have similar Values, the aim is to verify that the error on a new data (the *generalization error*) is of the same order of magnitude as LE and TE. We note in practice that an NN generalizes well if GE and LE are of the same order of magnitude and there are a priori bounds on the probability of the difference $|LE - TE|$ being larger than a certain value which we will discuss in Section 4.4. It is possible for some NN to give a bound on these errors, but there are still many open questions which we shall discuss in some detail in Section 4.4. Another important issue to examine is the *over-fitting* problem. If one deals with a time series of P data, the number of free parameters should be much less than the number of data; otherwise, the number of constraints will be equal or less than the number of parameters and so the probability of ending up with poor network performance becomes higher. Another important issue is that often the synaptic parameters realizing the minimum of LE (the learning error) do not necessarily realize the minimum of TE (the testing error). So the optimal choice is made by also looking at the behavior of TE. We will discuss these problems in detail in connection with our applications of Chapter 7.

4.2 Embedding Dimension

We shall review some methods for determining the embedding dimension of time series. Before going into the details we want to offer some general comments about the structure of the NN and the embedding dimension. We must bear in mind that the embedding dimension is the proper dimension of the input vector (see, e.g., Takens [27]) and that it coincides with the number N of input neurons. So the first layer of the NN is strictly connected with the structure of the time series or with the number of data of the series that are correlated. The number of neurons in the intermediate layer M is also connected with the data but it is found heuristically by optimizing the minimum learning error and is not directly connected with dependence properties within the time series. The two-layer NN is defined by the parameters N, M and by the synaptic connections between the two layers. It is not strictly necessary for all the neurons of the first layer to be connected with all the neurons of the second layer. The set of this information defines what we call *architecture* of the NN. We can also express the search for the embedding dimension by saying that we are going to look for the best architecture of the network. For practical reasons, we always have to look for architectures with a minimum number of neurons since it is difficult to handle an NN with many neurons due to the over-fitting problems which might arise. In particular, it is better to have as few neurons as possible because then we will have more vectors for the training and learning set. Once the architecture is fixed the process of learning and testing can be begun. During the testing one might decide to change the parameters a bit in order to obtain a smaller testing error. Concerning the number M, there are also some estimates for the two-layer NN which we will discuss in Section 5.1 and which can be useful for checking of the result obtained with these heuristic methods. We shall describe next the three main methods for finding the dependence among the data in order to also explain why we have chosen the correlation decay method for determining the embedding dimension.

4.2.1 Three Methods

In these subsections we summarize three methods for estimating the embedding dimension. The reason of our choice will be clear: the analysis of correlations is simpler and shorter than the other methods and thus this method is easier to use when dealing with a large mass of data as in our case or when the embedding dimension is rather large.

False neighbors. Suppose that we join the data of our time series in groups of m consecutive elements and we want to check whether m is the real embedding dimension of the data. This method is graphic in the sense that one plots the points defined by the m coordinates obtained with this procedure in an m-dimensional space. Then one starts labeling all the first neighbors for each point of the plot. If m is the real dimension of these data we should see that this set of neighbors also remains the same when we increase m. The justification of this fact is simple. Take a three-dimensional curve and project it onto a two-dimensional plane, the xy plane for example. Two points obtained in this way on the xy plane might be near even if their distance along the z axis is large. So if one includes the z coordinate in the set of coordinates for each point and looks at the plot of the representative points in the three-dimensional space one should observe a decrease in the number of neighbors for each point. Thus, the right embedding dimension is the one for which the increase of dimension leaves the number of neighbors unchanged. This method is interesting but it is difficult to realize for sets of vectors of large dimensions.

Relative entropy. If one considers the time series $X(t)$ as a stochastic process it is possible to evaluate with different methods the joint probability distribution $P_{tu}(x, y)$ defined as the density of the probability distribution of the event $X(t) = x$ and $X(u) = y$. Let $P_t(x)$, $P_u(y)$ be, respectively, the probability density of the events $X(t) = x$, $X(u) = y$. Then we can consider different entropies

$$
\begin{aligned}
S_{tu} &= -\int dx \int dy\, P_{tu}(x, y) \log P_{tu}(x, y), \\
S_t &= -\int dx\, P_t(x) \log P_t(x), \\
S_u &= -\int dx\, P_u(x) \log P_u(x).
\end{aligned}
\tag{4.8}
$$

The *relative entropy* is

$$
\Delta_{tu} = S_{tu} - S_t - S_u. \tag{4.9}
$$

It is clearly a measure of the probabilistic dependence of the variables $X(t)$ and $X(u)$. In fact, if these two variables are independent, then Δ_{tu} is equal to 0 for each t

and u; on the contrary, if they are dependent, the length of the interval $|t-u|$ such that $\Delta_{tu} = 0$ is a measure of the correlation length. If Δ_{tu} depends only on the distance of the points u and v then we have stationarity at the level of the two-point probability distribution of the time series. The relative entropy is a sensitive parameter which depends on the probability distribution of the joint events of the time series, but it is not very useful when dealing with large masses of data, as in our case, because it implies that one has to estimate the density of the probability distribution for each pair of data $X(t)$ and $X(u)$.

Decay of correlations. It is possible to use the following estimate for the correlation function of the time series $X(i)$:

$$r(k) = \frac{\frac{1}{P}\sum_{i=l}^{P-k}(X(i)-\bar{x})(X(i+k)-\bar{x})}{\frac{1}{P}\sum_{i=1}^{P}(X(i)-\bar{x})^2}, \tag{4.10}$$

where $\bar{x} = (1/N)\sum_{i=1}^{N} X(i)$. This way of writing the numerical approximation holds in the case of stationarity of the time series, otherwise r should depend on l and k.

The error of the approximation (4.11) can be estimated by use of the Bartlett formula ([10]):

$$\mathrm{var}(r(k)) \simeq \frac{1}{N}\left[1 + 2\sum_{v=1} r_v^2\right], \quad k > q. \tag{4.11}$$

The embedding dimension estimated with this method is simply the determination of k such that $|r(k)| \leq \mathrm{var}(r(k))$. In practice, if we have P data, then this approximation is meaningful for the first $P/2$ values of the index i. Due to the simplicity of this estimate, in Chapter 7 we have decided to use this method to analyze the embedding dimension of our data.

4.3 Learning Algorithms

There are two main algorithms used to minimize the error: steepest descent and simulated annealing.

4.3.1 Steepest Descent

This method finds a path in the space of parameters along which the function to be minimized has the maximum decrease. This path is parallel to the function's gradient. We explain how the method applies to the LE of an NN. We recall the formulas for the error function and the transformation of the NN. Using the same notation as in the first section, the output ξ^{μ} of the network is

$$\begin{aligned}
z_k^\mu &= \sum_{j=1}^{N} W_{kj} x_j^\mu, \\
h_k^\mu &= \sigma(z_k^\mu - \theta_k), \\
\xi^\mu &= \sigma\left(\sum_{k=1}^{M} V_{1k} h_k^\mu - \theta_0\right),
\end{aligned} \tag{4.12}$$

and the error, on the learning data set, is

$$LE = \frac{1}{P} \sum_{\mu=1}^{P} (\xi^\mu - y^\mu)^2. \tag{4.13}$$

We define by $\mathbf{W}$ the vector of all the weights $V_{11}, \ldots, V_{1M}$ and thresholds θ_j, $j = 1, \ldots, M$. The steepest descent applied to LE induces a transformation on the components of the vector $\mathbf{W}(n)$ into the components of a new vector $\mathbf{W(n+1)}$ where n denotes the number of iterations:

$$\mathbf{W}_j(n+1) = \mathbf{W}_j(n) - c(n) \frac{\partial LE}{\partial \mathbf{W}_j(n)}. \tag{4.14}$$

$c(n)$ is a decreasing function of n which must be chosen in such a way to guarantee the convergence of the algorithm to the absolute minimum $\mathbf{W}^*$ of LE as $n \to \infty$. It is evident that under such a transformation the function LE is decreasing. The choice of $c(n)$ might speed up the process of convergence, but this algorithm has a high probability of getting blocked in a local minimum. In fact, if the running point $\mathbf{W}(n)$ is located in a valley around a local minimum, it will inevitably approach this minimum and then remain blocked because there is no chance of increasing LE with this algorithm. The only way is to arbitrarily change the vector $\mathbf{W}(n)$ when the running point becomes blocked in the local minimum. A simple way to avoid this problem is to repeat the minimization procedure for various initial points and then to collect the values of the local minima. To avoid this lengthy procedure one should have a priori information about the behavior of LE. Usually this is not easy and so one must resort to a sequence of random trials or to an exploration of the entire phase space, which can be very long.

4.3.2 Simulated Annealing

In this section we use the symbol $\mathbf{W}$ with the same meaning as in the previous section. The alternative of starting from many random initial conditions for the learning algorithm is given by the minimization of the learning error function

$$LE = \frac{1}{P} \sum_{\mu=1}^{P} (\xi^\mu - y^\mu)^2 \tag{4.15}$$

by a stochastic dynamic applied to the space of configurations $\mathbf{W}$. This dynamic can be understood starting from the definition of the Markov chain defined by the

Monte-Carlo method ([20]). Let us consider a homogeneous Markov chain acting on the space of discrete vectors $\mathbf{W}(n)$ constructed in the following way. One starts with the hypothesis that all the components $\mathbf{W}_j$ vary in a bounded interval I, $I \equiv [-A, +A]$. The quantity A is estimated from the data and is usually of the same order of magnitude for all the components. Let us call Q the dimension of the vector $\mathbf{W}$, $Q = MN + M + 2$, where we suppose that there are only two thresholds, one for each layer, and that all the neurons belonging to the same layer have the same threshold. The interval I is divided in N_I small intervals of length $h, h = |I|/N_I$ and the transition from a configuration $\mathbf{W}(n)$ to the configuration $\mathbf{W}(n+1)$ takes place in the following way. First we assume that $\mathbf{W}(n+1)$ has only one component different from $\mathbf{W}(n)$, let it be the jth component. Then only the transition of this component of amplitude h is examined:

$$\mathbf{W}_j(n+1) = \mathbf{W}_j(n) \pm h.$$

We choose with probability $1/2$ the sign $+$ in front of h. Thus we have, for example,

$$\begin{aligned} \mathbf{W}_k(n+1) &= \mathbf{W}_k(n), \quad k \neq j, \\ \mathbf{W}_j(n+1) &= \mathbf{W}_j(n) + h. \end{aligned} \tag{4.16}$$

Then the new value of the error $LE(\mathbf{W}(n+1))$ is computed. If $LE(\mathbf{W}(n+1)) \leq LE(\mathbf{W}(n))$ then the new value $\mathbf{W}(n{+}1)$ of the weights is accepted. If $LE(\mathbf{W}(n{+}1)) \geq LE(\mathbf{W}(n))$ then the new value of the vector $\mathbf{W}$ is accepted with the probability ν

$$\nu = \frac{e^{-\beta \Delta LE}}{1 + e^{-\beta \Delta LE}}, \tag{4.17}$$

where ΔLE is the "energy" change

$$\Delta LE = LE(\mathbf{W}(n+1)) - LE(\mathbf{W}(n)).$$

This dynamic accepts with a certain probability transitions in which the "energy" LE increases. The parameter $\beta > 0$ is equal to the inverse "temperature" T of a system $\beta = 1/T$. Once we introduce the analogy of the error function LE with the energy of the system it is quite natural that the control parameter β is the inverse temperature. The temperature has the meaning of the average kinetic energy of the system and it is possible that for high temperatures one has higher jumps of the system. This is the interpretation of the formula for ν. If one increases T by a factor of ten, say T changes from 0.1 to 1, then one has that the possible jumps in energy ΔLE also increase by a factor 10 because ν remains the same. So this Markov chain, called also Monte-Carlo dynamics, allows the system to go out from local minima because there are also transitions in which the energy increases. It is clear that such transitions are controlled by β. One has to follow the evolution of the vector $\mathbf{W}$ and plot at each step n the value of the energy. If this value is not decreased, then an increase of the temperature is made in such a way that there is a possible increase of LE which brings the system out of the basin of attraction of the local minimum. Then the dynamic is started again, slowly decreasing the temperature of the system.

It is possible to show that the Monte-Carlo dynamics defined by a fixed value β is ergodic and the asymptotic probability distribution obtained by iterating the dynamic starting from an arbitrary initial value of $\mathbf{W}$ is the Gibbs distribution ([59]) with density $\rho(\mathbf{W})$:

$$\rho(\mathbf{W}) = \frac{1}{Z}\exp(-\beta E(\mathbf{W})),$$
$$Z = \int d\mathbf{W}\exp(-\beta E(\mathbf{W})). \tag{4.18}$$

From this formula it follows that the absolute minimum of LE is the configuration with highest probability for the Gibbs measure. Since the asymptotic values of $\mathbf{W}(n)$ are distributed with this measure, the event $\mathbf{W}(n) \to \mathbf{W}^*$ has the highest probability if one uses the Monte-Carlo method. But the Monte-Carlo method does not converge with probability 1 to the absolute minimum. An algorithm, called simulated annealing, which converges to the absolute minimum with probability 1 has been given in [20], [6]. In [20] the convergence of the algorithm to the absolute minimum with probability 1 is also proven. This theorem is the foundation of SA (simulated annealing). According to this result if β has a logarithmic increase with the index $l(n) = [n/M]$, $\beta(l(n)) = 1 + \log l(n)$, then the limiting configuration of the sequence $\mathbf{W}(n)$ is the absolute minimum of LE with probability 1. $l(n)$ is the number of times the algorithm examines the change of all the components of the vector $\mathbf{W}$, $[\ldots]$ means the integer part of the fraction. If the absolute minimum is degenerate then the limit configuration will coincide with one of the points realizing the absolute minimum. According to the above discussion the natural interpretation of the result of this proposition is that a logarithmic decrease of the temperature prevents the system from remaining blocked in local minima. The slow decrease of the temperature allows the system to make transitions with a higher energy increase, but with an amplitude decreasing with the order of iterations in such a way that the point does not remain blocked in unwanted "valleys." A graphical representation of simulated annealing is shown in Fig. 4.2.

We summarize this discussion with a scheme of application of MC (Monte-Carlo method) and SA (simulated annealing).

- $\mathbf{W}(n) \in \Gamma = [-A, A]^Q$. The configuration vector of the system will always remain in this cube of dimension Q.
- The intervals $I = [-A, A]$ are divided in N_I intervals of length h

$$h = \frac{|I|}{N_I} \tag{4.19}$$

  where h is the elementary step of the algorithm.
- The components of the configuration vector $\mathbf{W}$ are the synaptic weights and the thresholds θ of the NN:

$W_{1,1}$	...	$W_{M,N}$	V_1	...	V_M	θ_0	θ_1

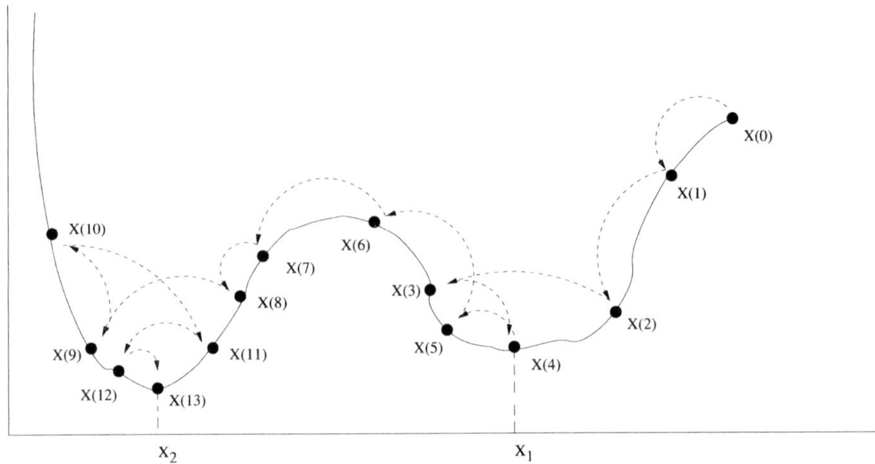

Fig. 4.2. Trajectory of simulated annealing.

- The initial values of the components of the vector **W** are chosen with uniform probability distribution in Γ. Usually the values of the thresholds θ are chosen from a certain interval before starting the dynamics. For each choice the dynamic is repeated until the SA converges to the absolute minimum. In this way the simulated annealing acts only on the space of the synaptic weights.
- If the output of the value of the random subroutine belongs to the interval $[j/Q, (j+1)/Q]$ then MC is applied to the jth component of **W** in the sense explained below.
- If the output of the random subroutine is less than $1/2$ then $\mathbf{W}_j(n+1) = \mathbf{W}_j(n) + h$, i.e., the increase in the jth component of the vector $\mathbf{W}(n)$ is considered, otherwise the increment of opposite sign is examined: $\mathbf{W}_j(n+1) = \mathbf{W}_j(n) - h$.
- One computes $\Delta LE = LE(\mathbf{W}(n+1)) - LE(\mathbf{W}(n))$.
  Then the transition from $\mathbf{W}(n)$ to $\mathbf{W}(n+1)$ takes place in this way:
  a. If $\Delta LE < 0$ the new vector is accepted: $\mathbf{W}(n) \to \mathbf{W}(n+1)$
  b. If $\Delta LE \geq 0$, then $\mathbf{W}(n) \to \mathbf{W}(n+1)$ if rand(0) $\leq \nu$ with ν defined by

$$\nu = \frac{e^{-\beta \Delta LE}}{1 + e^{-\beta \Delta LE}}. \tag{4.20}$$

- In order to implement the SA we introduce a dependence of β on $l(n)$:

$$\beta(n) = 1 + \log l(n)$$

  where $l(n)$ is the number of times the MC explores all the components of **W**. The initial value of β is chosen with preliminary trials before starting the SA.
- SA is repeated for a sufficiently large number of n until the learning error is smaller than at least one order of magnitude of the average values of the data.

When the learning error is smaller than the above-mentioned value, the test procedure is started and, if the testing error remains of the same magnitude, the NN is ready for its operation of forecasting or reconstruction of missing data.

4.4 Rigorous Results

There are many papers, published in particular in journals concerned with applications, where NNs are used to analyze or predict time series. Two such fields are meteorology and hydrology. The estimate of the error relative to the network's performance is usually done numerically on the set of data while there is almost no discussion of the problem of the network's generalization ability. The problem of the analytical estimate of the prediction error for classical statistical models has been investigated systematically, but the absence of references to the analogous problem for NNs may suggest that this question has not been treated for the NN. This impression is false—there is a lot of work going on in this direction. In this section we explain the main rigorous estimates on the learning curve for NNs. We also mention what is really still at the level of conjecture: if the time series is long enough, there is a large probability that GE and LE have almost coinciding values. In fact, this is still a conjecture for general NNs and has been proven with various methods only for special cases, which are discussed in the following sections. Also, the methods for proving such an important fact differ greatly in various cases so at present the hope for a proof for the general case is small. The estimate of the learning and generalization errors has already been analyzed in many papers, and we can divide the different approaches into two main types. One considers the NN as a particular algorithm for approximating some unknown function which generates the data. We call this approach *approximation theory* and we describe some of these results in Chapter 5. The other approach is mainly probabilistic and is based on estimates from above of the probability that the learning error and the generalization error differ by more than a small constant. We call this the *probabilistic approach* and we describe it in the next subsection. These two approaches are quite different and yet the information they provide is almost complementary so we need to be acquainted with both of them. The main difference between these two methods is that in the second one the NN for a specific time series is constructed by a minimization procedure while in the first one the NN is given by a direct computation from the data of the time series. We prefer the second method because in our applications it has been found to be more effective as we will show in Chapter 8. The first method requires much less work and has more rigorous estimates than the second so it is more attractive for a theoretical point of view. However according to our results, it is not always effective in practice.

4.4.1 Vapnik–Chervonenkis Theorem

One of the first important estimates was given in the seventies by Vapnik and Chervonenkis ([64]). To be precise, we should note that the authors derived a general property of the law of large numbers that had been applied to the theory of NN only after many years. The theorem shown below solves the problem of the uniform convergence of the frequencies of the events belonging to a certain class Ω to their probabilities. It is a well-known fact that under the assumptions of the law of large numbers the frequency of an event $\omega \in \Omega$ converges in probability to the probability of ω, but from this fact it does not follow at all that the frequencies of all the events in Ω converge *uniformly* in Ω to the corresponding probabilities. We show the theorem of Vapnik and Chervonenkis for the case of NNs, although it can be applied directly only in the case of binary inputs and outputs. We define as a *sample* the sequence of pairs of input-output patterns in the learning set $\{\underline{x}^\mu, y^\mu\}_{\mu=1}^P$. Let Ω be the compact set to which all the pairs belong. The learning error coincides with the frequency of the event such that the network, applied to the input $\underline{x}^\mu$, differs from the output y^μ by an amount less than a certain $\epsilon > 0$. The main task is to estimate the probability that the output of the network z^μ is different from the real value y^μ for any pair $\{\underline{x}, y\} \in \Omega$. Then the statement of the theorem ([66]) is as follows:

Theorem 4.1 *For any $\eta > 0$ and P large enough there is a positive constant $m(2P)$ such that*

$$\sup_{\xi \in \Omega} P(|\nu_\xi - P_\xi| \geq \eta) \leq m(2P) \exp(-\eta^2 P/8), \tag{4.21}$$

where ν_ξ is the frequency of the error for a single event $\xi^\mu = (\underline{x}^\mu, y^\mu)$ of the sample $\xi = (\xi^1, \ldots, \xi^P)$ while P_ξ is the probability of the error on the pattern in Ω.

The meaning of this theorem in the language of NNs is that the probability that the generalization error differs from the learning error by a quantity bigger than $\eta > 0$, for any input belonging to a bounded set Ω, decays exponentially with the length of the sample P (i.e., with the number of elements of the learning set) multiplied by a constant $m(2P)$ depending only on P. This happens uniformly in ξ, i.e., the constants do not depend on $\xi \in \Omega$. It is not surprising that this neat proposition cannot be applied to any kind of NN since it is very strong and does not depend on the rule that generates the data. As we will see in the following examples, all the estimates of the probability of having a given difference among LE and GE are made starting from the knowledge of the real rule that generates the data, since this generating rule allows estimates of the asymptotic forms of the generalization and learning errors. The constant $m(2P)$ has a combinatorial meaning and it might be bounded from above by $\exp \mathrm{const}\, P$ for $P \leq d_{VC}$ and by P^n for $P \geq d_{VC}$, where n is a positive integer evaluated during the proof of the theorem (4.21) and d_{VC} is a typical constant of the data set and is called the Vapnik–Chervonenkis dimension. In [65] an estimate of this type has been made only for the *learning machine*, which is a network acting on vectors with components ± 1 and with binary outputs, thus a network acting only on patterns of binary form. A classical example of a learning machine is the linear perceptron which fulfills the XOR rule among patterns of the form (n, m) with $n, m = 0, 1$. Another example is the support vector machine discussed in the next section. The problem of generalization to any type of patterns and networks is still not fully understood. It is possible to consider the variables of the problem on a lattice and use this result, but then one should find estimates uniform with respect to the lattice step. Thus, we can conclude from Theorem 4.1 that the probability that the generalization error differs from the learning error in a sample of length P decays exponentially with P if $P \geq d_{VC}$. If one has a reasonable estimate of d_{VC} then the use of this theorem is straightforward, because one has the a priori estimate which shows that if the learning error of an NN is 5% (for example) then the probability that the generalization error is larger than $\eta + 5\%$ is less than $\exp -(\eta^2)P/8$ if $P \geq d_{VC}$. Thus, if one knows d_{VC} it is possible to use the exponential decay in the probability estimates and in order to have a generalization error less than 1% we find that $P > 20{,}000$ if $20{,}000 \geq d_{VC}$. Thus, efforts are concentrated on the problem of the estimate of d_{VC}, which is still an open question. We refer to the literature mentioned above for documentation on this problem or also on some interesting review papers.

4.4.2 Support Vector Machine

We should note that the estimate of the generalization error has also been investigated with other techniques. The problem of estimating the generalization error is somehow different because the question is about the probability of guessing from the learning set a good approximation of the function $f(\underline{x}^\mu)$ which generates the data. As we mentioned, it is necessary to consider the function generating the data as known in order to obtain adequate estimates (probabilistic or not) of LE and GE. In other words, we suppose that there exists such a function, as we did in Section 4.1, and we want to have an estimate of the probability that the guess of $f(\underline{x}^\mu)$ is the right one. Here $\underline{x}^\mu$ is the usual N-dimensional vector of real numbers. These studies have been done starting from the simplest case: the support vector machine ([18]). The support vector machine is used for classification of data into two different classes. Suppose that each class corresponds to a given half-space defined by a certain hyperplane in the n-dimensional space of patterns. The classification rule is that each pattern is a member of the ± 1 class according to which half-space it belongs. It is easy to represent this rule using a non-linear preceptron, i.e., an NN with no hidden layer. The approach currently used for finding the classification rule from a sequence of input-ouput patterns is as follows: suppose that there is a non-linear perceptron generating the pairs of $\underline{x}^\mu$, ξ^μ of input ($\underline{x}^\mu$) and output ξ^μ data defined by the vector of synaptic weights $\underline{w}_0$. The function which associates the output to the input

$$y = \text{sign}((\underline{w}_0, \underline{x})) \tag{4.22}$$

is called the *teacher rule* and is defined by the vector $\underline{w}_0$. The symbol $(\underline{x}, \underline{y})$ indicates the scalar product among the vectors $\underline{x}$ and $\underline{y}$ and sign(u) is the sign of the variable u. The interpretation of this rule is straightforward. It classifies the points of R^N by means of the hyperplane passing through the origin and defined by the vector $\underline{w}_0$ perpendicular to it. All the points which have a coordinate vector forming an angle with $\underline{w}_0$ belonging to the interval $(-\pi/2, +\pi/2)$ are classified with a $+$ by means of the scalar product. All the points lying on the other half-space have a minus sign. Suppose that we have a sequence of patterns $\underline{x}^\mu$, ξ^μ, $\mu = 1, \ldots, P$ generated with the teacher rule; then the "student" will guess the rule knowing that it is of the type (4.22)

$$y = \text{sign}((\underline{w}, \underline{x})) \tag{4.23}$$

and arguing on the basis of the patterns $\underline{x}^\mu, \xi^\mu$ in this way. Suppose that the first point $\underline{x}^1$ has a signature $+$. Then $\underline{w}_0$ should make an angle of less than 90 degrees with the vector $\underline{x}^1$ and so it should belong to the same half-space defined by the plane $(\underline{x}^1, \underline{z}) = 0$ where the vector $\underline{x}^1$ lies. The next pattern of the learning set $\underline{x}^2, \xi^2$ defines another half-space of vectors in the same way. Suppose that $\xi^2 = -1$, then the vector $\underline{w}_0$ should belong to the half-space of vectors having an obtuse angle with the vector $\underline{x}^2$. Then from the two patterns the "student" can guess that the unknown vector $\underline{w}_0$ belongs to the intersection of these two half-spaces. If the "student" has only these two patterns he can guess that the rule is of the type (4.23) with a vector $\underline{w}$ belonging to this intersection. If α_2 is the angle defined by the two hyperplanes, clearly the error of determination of $\underline{w}_0$ is given by α_2 and so the probability of wrong generalization of the student rule is $\epsilon_2 = \alpha_2/2\pi$. The question which has been investigated is to determine the limit of LE for $P \to \infty$. This problem has been solved with a statistical mechanics approach which we summarize here ([21]). We choose for simplicity the vectors $\underline{x}^\mu$ to have only Bernoulli independent components ± 1 and we call them $\underline{\xi}^\mu$, the output is ζ^μ. Given a sequence of patterns $\underline{\xi}^\mu, \zeta^\mu, \mu = 1, \ldots, P$, the learning error LE is a function of the sequence of patterns $\xi \equiv \underline{\xi}^\mu, \zeta^\mu$, $\mu = 1, \ldots, P$ of the vector guessed by the student $\underline{w}$ and the vector of the teacher $\underline{w}_0$ and so we write it as $LE(\xi, \underline{w}, \underline{w}_0)$. Then we can write

$$LE(\xi, \underline{w}, \underline{w}_0) = \frac{1}{P} \sum_\mu \Theta(-(\underline{w}, \underline{\xi}^\mu)(\underline{w}_0, \underline{\xi}^\mu)) \tag{4.24}$$

where $\Theta(z)$ is a function which equals 1 for $z \geq 0$ and equals 0 for z negative. The contributions to the above sum are different from zero only when the sign of the scalar product of the "teacher" rule $(\underline{w}, \underline{\xi}^\mu)$ differs from the sign of the scalar product of the "student" rule $(\underline{w}_0, \underline{\xi}^\mu)$ which is equivalent to the occurrence of an error in the learning set. Let us suppose that the minimum of the learning error is searched by means of the Monte-Carlo method. In this particular case the Monte-Carlo method is a sequence of vectors $\underline{w}$ generated by a Markov chain with probability transition defined in (4.20) where the energy E is LE (4.24). With this procedure a sequence of weight vectors is created $\underline{w}(1), \ldots, \underline{w}(n)$ which converges with probability one, with respect to a random choice of the initial value $\underline{w}(0)$, to the absolute minima $\underline{w}^*$ of the error. For vectors of $\underline{w}$ with a finite number of components the probability transition of this particular Monte-Carlo dynamic is bounded from below and so we have that the Monte-Carlo dynamic is an ergodic Markov chain (see Gnedenko [67]). From the ergodicity of the Monte-Carlo dynamics it follows that

$$\lim_{T\to\infty} \frac{1}{T} \sum_{n=1}^{T} LE(\xi, \underline{w}(n), \underline{w}_0) = \int d\mu_G(\xi, \underline{w}, \underline{w}_0) LE(\xi, \underline{w}, \underline{w}_0), \tag{4.25}$$

where we have introduced the Gibbs measure μ_G, defined as follows.

Definition 4.1

$$d\mu_G(\xi, \underline{w}, \underline{w}_0) \equiv d\mu(\underline{w}) \frac{\exp -\beta PLE(\xi, \underline{w}, \underline{w}_0)}{Z(\xi, \underline{w}, \underline{w}_0)} \tag{4.26}$$

is the Gibbs measure generated by the learning error $LE(\xi, \underline{w}, \underline{w}_0)$. $d\mu(\underline{w})$ is the Riemann measure on the N-dimensional space of the weight vectors $\underline{w}$ and Z is the partition function:

$$Z \equiv \int d\mu(\underline{w}) \exp(-\beta PLE(\xi, \underline{w}, \underline{w}_0)).$$

The parameter $\beta > 0$ has the meaning of inverse temperature for systems of statistical mechanics. In this case it represents the internal noise of the system's neurons. This noise interferes with the generalization ability of either the "teacher" or "student" rule. It is natural to represent this noise as a temperature. The equality in (4.25) is a consequence of the ergodicity of the Monte-Carlo method and of the fact that the Gibbs measure (4.26) is invariant with respect to the Monte-Carlo dynamic (see Tirozzi [59]). This equality is of fundamental importance because it connects computer simulations (Monte-Carlo) to statistical mechanics averages with respect to the Gibbs measure $d\mu_G(\xi, \underline{w}, w_0)$ and vice versa. Let us also note that the left-hand side of (4.25) converges to the value of the absolute minima of the learning error $LE(\xi, \underline{w}^*, \underline{w}_0)$ since we suppose that MC will bring the weights to the absolute minimum. Also significant is the fact that the convergence to the absolute minimum is proved only in the framework of SA but cannot be proven for the MC method. Thus, from the hypothesis of convergence of the MC to the absolute minimum it follows that the learning error is given by the average with respect to the Gibbs measure defined on the right-hand side of (4.25). From these arguments it follows that

$$LE(\xi, \underline{w}^*, \underline{w}_0) = \int d\mu_G(\xi, \underline{w}, \underline{w}_0) LE(\xi, \underline{w}, \underline{w}_0), \tag{4.27}$$

which can be written as

$$LE(\xi, \underline{w}, \underline{w}_0) = -\frac{1}{P}\frac{\partial}{\partial \beta} \log Z(\xi, \underline{w}, \underline{w}_0). \tag{4.28}$$

For a finite number of patterns and for vectors $\underline{w}$ with a finite number of components the free energy function Φ

$$\Phi \equiv -\frac{1}{P} \log Z(\xi, \underline{w}, \underline{w}_0)$$

is analytic with respect to β. Difficulties arise for large values of the number of components of the state vector $\underline{w}$ or for large values of the number of patterns P.

4.4.3 Statistical Mechanics Approach

The concepts introduced in the previous section originate from the statistical mechanics methods which have been used extensively in many areas of physics and have been usefully applied in studies on the generalization ability of NNs. The main fact is that all the neat properties of retrieval and generalization hold for large values of the number of patterns, but in this case more complex arguments of probability theory should be used. In fact, $LE(\xi, \underline{w}, \underline{w}_0)$ depends on many random variables: the sequence ξ of independently chosen random patterns, the "student" rule $\underline{w}$ and the "teacher rule" $\underline{w}_0$. The same happens for the generalization error

$$GE(\xi', \underline{w}, \underline{w}_0) = \Theta(-(\underline{w}, \underline{\xi}')(\underline{w}_0, \underline{\xi}')), \tag{4.29}$$

where ξ' is a new incoming pattern not belonging to the learning set ξ. The interesting case is when there is independence of the learning error and generalization error from the sequence of patterns ξ and from the "teacher" rule at least. This is possible only in the limit $P \to \infty$ and we can express this property in more exact mathematical terms using the concept of *self-averaging*. Let us call E_ξ the expectation with respect to the probability of the sequence of patterns. The form of this probability is simple because the ξ are independent random variables with values ± 1. Thus, this average is defined as

$$E_\xi(F(\xi)) = \frac{1}{2^{NP}} \sum_{\xi_i^\mu = \pm 1} F(\xi). \tag{4.30}$$

We can express the asymptotic independence of LE and GE from these random parameters using the self-averaging property:

Definition 4.2 *Any function $F(\xi, \underline{w})$ is self-averaging if*

$$\lim_{P \to \infty} (F(\xi, \underline{w}) - E_{\xi \underline{w}} F(\xi, \underline{w}))^2 = 0. \tag{4.31}$$

This condition is equivalent to the convergence in probability of $F(\xi, \underline{w})$ to its average with respect to ξ and $\underline{w}$. The symbol $E_{\xi \underline{w}}$ is the average with respect to ξ and $\underline{w}$:

$$E_{\xi \underline{w}} = E_\xi \int d\mu(\underline{w}).$$

Thus, the ability of an NN to generalize well can be derived starting from the self-averaging property, which must be shown for each particular case. There is, however, no proof yet of such a fact for the particular NN we are considering now, while for general back-propagation NNs, this property must be checked numerically. We conclude the description of the non-linear perceptron by showing the method used by physicists for computing the free energy of the system for large values of P. This method is called the *replica trick*, and it is not rigorous. Moreover, there is as yet no proof of the self-averaging property for the non-linear perceptron. We merely mention the replica trick procedure and give the result. Starting from (4.28) we need to compute

$$-\frac{1}{P} \frac{\partial}{\partial \beta} E_{\xi \underline{w}} \log Z(\xi, \underline{w}, \underline{w}_0). \tag{4.32}$$

The replica trick consists of exchanging the expectation with the logarithm

$$E_{\xi \underline{w}} \log Z(\xi, \underline{w}, \underline{w}_0) = \lim_{x \to 0} \frac{E_{\xi \underline{w}} Z(\xi, \underline{w}, \underline{w}_0)^x - 1}{x}. \tag{4.33}$$

The expectation $E_{\xi \underline{w}} Z(\xi, \underline{w}, \underline{w}_0)^x$ can be computed only for integers x, say n. After an intriguing and long calculation (see the book of Mezard, Parisi and Virasoro [68]), the limit $x \to 0$ is done sending $n \to 0$ in the obtained expression. This trick is non-rigorous because there is no justification that it is possible to make an analytic continuation from integer n to real x, nor is there a proof that such a limit exists and that it is unique. Nevertheless, the replica trick has been fruitful for obtaining many interesting results. The final result given by the heuristic replica trick for the expression of the GE for the non-linear perceptron is

$$GE(\xi) = \frac{1}{\pi} \int d\mu(\underline{w}) \arctan((\underline{w}_0, \underline{w})). \tag{4.34}$$

The central limit theorem can be used in the case of some special models to evaluate the learning and generalization errors. We now describe the case of the Hebbian learning rule (Vallet [63]). Let us take all the $\underline{x}^\mu$ vectors of the learning set to have components ± 1 as before. The output ζ^μ of the teacher rule is given by

$$\zeta^\mu = \operatorname{sign}((\underline{w}_0, \underline{\xi}^\mu)). \tag{4.35}$$

The Hebb rule defines a vector $(\underline{w})$ associated to the sequence of patterns $(\underline{\xi}^\mu, \zeta^\mu)$:

$$w_i = \frac{1}{\sqrt{\alpha N}} \sum_{\mu=1}^{P} \operatorname{sign}((\underline{w}_0, \underline{\xi}^\mu)) \xi_i^\mu, \tag{4.36}$$

where $i = 1, \ldots, N$, and $\alpha = P/N$ is the capacity of the network. The Hebb rule gives an estimate of the weight vector by a formula based on the knowledge of the input vectors ξ^μ and the corresponding output ζ^μ. The factor $1/\sqrt{\alpha N}$ in the above formula derives from the normalization condition $(\underline{w}, \underline{w}) = 1$ and the fact that the patterns $\underline{\xi}^\mu$ are also normalized:

$$\sum_i (\underline{\xi}^\mu)_i^2 = 1.$$

Then the non-linear perceptron with the weights (4.36) generalizes well if for any vector $\underline{\xi}$ it happens that $(\underline{\xi}, \underline{w})(\underline{\xi}, \underline{w}_0) \geq 0$. Substituting the definition we find that the probability of wrong generalization for a generic incoming vector $\underline{\xi}$ is

$$GE(\underline{\xi}) = \text{Probability} \left\{ \sum_{\mu=1}^{P} \operatorname{sign}((\underline{w}_0, \underline{\xi}^\mu))(\underline{\xi}^\mu, \underline{\xi}) \operatorname{sign}((\underline{w}_0, \underline{\xi})) \leq 0 \right\}. \tag{4.37}$$

The generalization error ϵ is then the average of $GE(\underline{\xi})$ over all the vectors $\underline{\xi}$ chosen with uniform probability:

$$GE = \frac{1}{2^N} \sum_{\underline{\xi}} GE(\underline{\xi}). \tag{4.38}$$

If one chooses $\underline{\xi} = \underline{\xi}^1$ then one obtains the learning error LE:

$$LE(\underline{\xi}^1) = \text{Probability} \left\{ 1 + \sum_{\mu=2}^{P} \text{sign}((\underline{w}_0, \underline{\xi}^\mu))(\underline{\xi}^\mu, \underline{\xi}^1)\text{sign}((\underline{w}_0, \underline{\xi}^1)) \leq 0 \right\}. \tag{4.39}$$

Taking the limit $P \to \infty$, $N \to \infty$, $P/N \to \alpha$ it is possible to apply the central limit theorem and obtain the asymptotic form of LE and GE ([63]):

$$\begin{aligned} LE &= \frac{1}{\sqrt{\pi}} \int_0^\infty \exp(-u^2) \text{erfc} \left(-u\sqrt{\frac{2\alpha}{\pi}} - \frac{1}{\sqrt{\alpha}} \right) du, \\ \text{erfc}(x) &= \frac{2}{\sqrt{\pi}} \int_x^\infty du \exp(-u^2), \\ GE &= 1 - \frac{1}{\sqrt{\pi}} \text{arctg} \sqrt{\frac{\pi}{2\alpha}}. \end{aligned} \tag{4.40}$$

We notice that the result is independent of the specific choice of $\underline{w}_0$ after making the average with respect to $\underline{\xi}$. This is a consequence of the general property of this particular NN (Hopfield type with Hebb's weight vectors) which is nothing else than the *self-averaging* property introduced before. It expresses the fact that asymptotically the average performance of the network, for the particular case of Hebb's rule, is independent of the initial choice of the synaptic weights and so one can compute these errors averaging with respect to the synaptic vectors. Without this property, one cannot estimate the average, and the evaluation of these important quantities would be impossible. The consequence of this result is that for increasing α the generalization error and learning error tend to the same quantity in accordance with the general idea of NNs. However, this argument, as well as the Vapnik–Chervonenkis theorem, also shows that it is not easy to prove such a property in the general case. Amari ([7]) has an upper bound for the error of the type const$/P$ using the Bayesian formulation of the probability of the good generalization of the non-linear perceptron. In his proof, though, he assumes some regularity properties of the data which are difficult to show in concrete cases. Furthermore, a velocity of convergence of the partition function to a deterministic limit is assumed, but not shown. In Amari's proof, the partition function is defined as the probability that the output of the network is correct for a given sample. It is important to show the convergence for $N \to \infty$ (where N is the number of input neurons) of the partition function to some constant and, in fact, this is a property of self-averaging which also must be shown. Moreover, for the case of the Hebb's perceptron considered above, this property is not true. The velocity of approach of the Amari partition function to a limit is a result of large deviations which have not been proven in this case. Interesting results of this type have been obtained using methods of statistical mechanics ([21], [58], [57]) when the teacher rule is influenced by a noise. We shall briefly discuss the results of [58], [57] and their generalization made by Shcherbina and Tirozzi ([54]).

In [58], [57] only the case of a linear perceptron as a "teacher" rule

$$y = (\underline{w}_0 \underline{x}) \tag{4.41}$$

is considered, whereas in [54] analogous results are shown for the more realistic case of the non-linear perceptron:

$$y = \sigma(\underline{w}_0 \underline{x}). \tag{4.42}$$

Furthermore, in [54] a rigorous proof has been given which avoids the non-rigorous replica calculations of [58], [57]. We start by taking the "teacher rule" to be a non-linear perceptron defined by the vector of synaptic weights $\underline{w}_0$ summed to a Gaussian noise:

$$y^\mu = \sigma(\underline{w}_0 \underline{x}^\mu) + t\eta^\mu, \tag{4.43}$$

where η^μ is a Gaussian noise. One looks for the best approximating "student rule"

$$z^\mu = \sigma(\underline{w}\underline{x}^\mu) \tag{4.44}$$

in the usual sense where the learning error must be minimum on the data set $(\underline{x}^\mu, y^\mu)_{\mu=1}^P$:

$$LE(t) = \frac{1}{P}\left\langle\left\langle\left\langle \sum_{i=1}^{P} (\sigma(\underline{w}_0, \underline{x}^i) - \sigma(\underline{w}, \underline{x}^i) + t\eta^i)^2 \right\rangle\right\rangle\right\rangle_\eta, \tag{4.45}$$

where $\langle\ \rangle$ is the average with respect to the Gibbs measure defined on R^n, the space of the synaptic weights $\underline{w}$. Here $\langle\ \ \rangle_\eta$ is the average with respect to the random variables $\eta^1, \dots, \eta^P$. Thus,

$$\langle f(\underline{w}) \rangle = \int_{R^n} d\underline{w}\rho(\underline{w}) \frac{1}{Z_P(\beta, x, \eta)} f(\underline{w}) \exp(-\beta E(\underline{w}, x, \eta)) \tag{4.46}$$

where ρ is a probability distribution defined on R^n, for example, a $N(0, 1)$ Gaussian, and $E(\underline{w}, x, \eta)$ is the sum appearing in (4.45). It is a measure of the squared difference between the "teacher rule" and "student rule" evaluated on the input data $x = (\underline{x}^1, \dots, \underline{x}^P)$ of the learning set. These data are influenced by a Gaussian noise $\eta = (\eta^1, \dots, \eta^P)$, the η^μ being independent random variables. The function $E(\underline{w}, x, \eta)$ is considered to be the Hamiltonian of the equivalent statistical mechanics problem. Its form is equivalent to the learning error LE (4.2) where y^μ is given by (4.43) and z^μ by (4.44):

$$H_P(\underline{w}, x, \eta) = (1/2P) \sum_{i=1}^{P} (\sigma(\underline{w}_0, \underline{x}^i) - \sigma(\underline{w}, \underline{x}^i) + t\eta^i)^2. \tag{4.47}$$

The parameter t measures the intensity of the external noise on the data. The sample variables $\underline{x}^i$ are extracted with a certain probability distribution from the space of data and so they must also be considered as independent Gaussian random variables. Z_P is the normalization factor of the Gibbs measure and is the partition function of statistical mechanics introduced before:

$$Z_P(\beta, x, \eta) = \int_{R^n} d\underline{w}\rho(\underline{w}) \exp(-\beta LE(\underline{w}, x, \eta)). \tag{4.48}$$

The generalization error can be evaluated by exploiting the equivalence of the statistical mechanics approach and the Bayesian approach ([57], [7]):

$$GE(t) = \left\langle -\log \frac{Z_{P+1}}{Z_P} \right\rangle_\eta . \tag{4.49}$$

In [54] it has been shown that $LE(t)$, $GE(t)$, for asymptotical large values of P, decrease in a way inversely proportional to P if there is no noise in the network. The main tools used in the proof have been the saddle point method and the self-averaging property with respect to the random choices of the patterns x. Thus, the same results obtained in [58] for the case of the linear perceptron by means of the non-rigorous replica trick are obtained here in a rigorous way for the more general case of the non-linear perceptron. We can then write the following theorem ([54]):

Theorem 4.2 *Under the above hypotheses for* $P \to \infty$,

$$E\{LE(0)\} = \frac{n}{2\beta P} + o(P^{-1}), \quad E\{GE(0)\} = \frac{n\beta}{2P} + o(P^{-1}). \tag{4.50}$$

$$E\{LE(t)\} = \frac{t^2}{2} + O(P^{-1}), \quad E\{GE(t)\} = \frac{\beta t^2}{2} + O(P^{-1}). \tag{4.51}$$

The proportionality constant is given by the dimension n of the input vector in the case of zero external noise. If there is an external noise, the error of generalization and learning cannot be less than $t^2/2$, as is easy to understand from this theorem. Thus LE and GE decrease asymptotically in the same way and this is agrees with the estimates of Amari and Feng discussed in this section. Such a property is a good basis for believing the Vapnik and Chervonenkis result for the case of continuous variables, but without the exponential decay of the generalization error. Unfortunately, it is difficult to extend these neat results to the case of a two-layered NN; indeed, this does not yet seem to have been done.

4.4.4 Extreme-Value Theory

Feng ([18]) has used an original approach using the extreme-value theory of statistics ([34]) for an asymptotic estimate of the generalization error. Suppose that the data

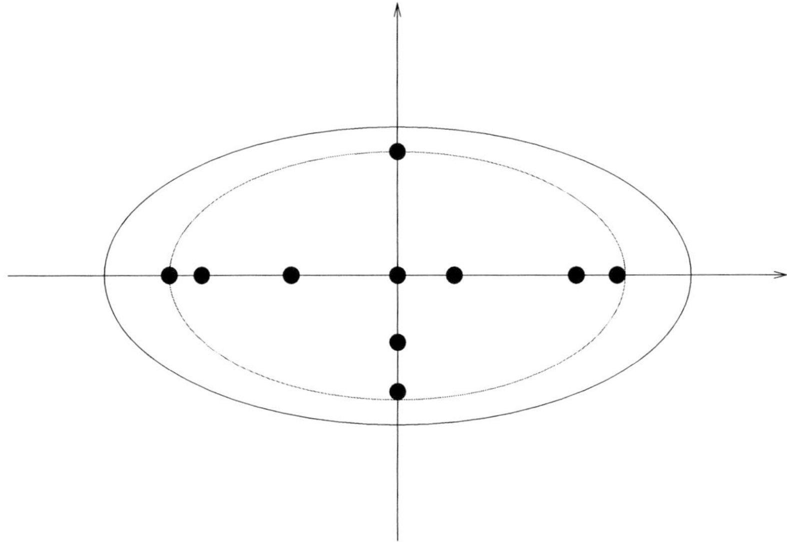

Fig. 4.3. Generalization error for the perceptron.

are two dimensional (x, y) and that the "teacher" rule assigns $+1$ to the points of the plane lying outside an ellipse of semiaxis a and b lying, respectively, on the x axis and on the y axis and -1 corresponds to the points inside the ellipse. Suppose also that the sample points are of the type $(x_1, 0)$ or $(x_2, 0)$: the student will observe M of such points with their corresponding sequence of $+1$ and -1. Let N be the subset of the sample points with associated value -1.

Then the student will conclude that the points inside the ellipse with semiaxes

$$a_N = \max(x_1, \ldots, x_N) \tag{4.52}$$

$$b_N = \max(y_1, \ldots, y_N) \tag{4.53}$$

have -1 as output. The generalization error is thus the area of the true ellipse minus the area of the ellipse with semiaxis a_N, b_N (see Fig. 4.3):

$$GE = \pi(ab - a_N b_N).$$

From extreme-value theory we find that $a_N \to a(1 - 1/N)$, $b_N \to b(1 - 1/N)$ as $N \to \infty$ since the random variables $x_1, \ldots, x_N$ and $y_1, \ldots, y_N$ are independent. Thus, the generalization error decreases as $1/N$.

Unfortunately, it is not simple to extend this interesting and elegant method to a multilayer perceptron and to the case of a sigmoidal input-output function.

5

Approximation Theory

In this chapter we discuss and show some results for the use of the neural network (NN) as a complete set of functions. The fact that the combination of the sigmoidal function corresponding to an NN can approximate any function is a simple consequence of the Stone–Weierstrass theorem and so such an approach is a convincing one. Furthermore, in the case of approximation theory the synaptic weights are given by some a priori estimates and in many cases could be directly evaluated from the data. This approach has, as a drawback, more errors than the NN constructed using the procedures described in the previous chapter.

5.1 General Position of the Problem

Since the approximation of any function generating the time series is the central issue of the NN algorithm, it is important to consider how the mathematical literature deals with the problem of approximating any reasonable function with a complete set of functions. The sigma functions with an argument given by linear combinations of the input data are a particular case of a complete system. So given a general class of functions $f(\underline{x})$, we look for its approximation with a linear combination of the type

$$\sum_{k=1}^{l} v_k \sigma \left(\sum_{j=1}^{n} w_{kj} x_k - \theta_k \right). \tag{5.1}$$

The architecture corresponding to this particular choice is a two-layer back-propagation (see Chapter 4) with the difference that the output neuron is linear. There are general theorems about the generalization error obtainable by the application of this structure: here we show one of them. The importance of this theorem is that it gives a motivation for the application of NNs of this type.

Theorem 5.1 *Let $f(\underline{x})$ be a continuous function defined in a compact subset K of R^n. For any $\epsilon > 0$ there exist:*

- *an integer l*

- *real numbers* w_{kj}, $1 \le k \le l$, $1 \le j \le n$
- θ_k *real* , $1 \le k \le l$
- v_k *real,* $1 \le k \le l$

such that for any $\underline{x} \in K$*:*

$$\left| f(\underline{x}) - \sum_{k=1}^{l} v_k \sigma \left(\sum_{j=1}^{n} w_{kj} x_j - \theta_k \right) \right| \le \epsilon. \tag{5.2}$$

If one looks for an accuracy of the order of ϵ it is possible to find a network with l neurons in the hidden layer and with the particular structure described above which gives an approximation less than ϵ uniformly on a compact K. In other words, for any point $\underline{x} \in K$ the error made by approximating the unknown function $f(\underline{x})$ with the NN of this type is the same. There are theorems which show that l must be larger than or equal to ϵ^{-n} and so the complexity of the network increases with the power of the inverse of the required accuracy. This estimate is uniform in K and needs a further hypothesis: the supremum norm (see 5.26) of the gradient of $f(\underline{x})$ must be less than 1 ([42]). The proof of this theorem can be found in [12], [13], [38], [43], [44]. This structure of the NN and this methodology will be used in the particular case shown in the following sections with a certain choice of the coefficients. The application is discussed in Chapter 8.

5.2 Explanation of the Mathematics

Before explaining the mathematical construction of the next section, we would like to present the main ideas and the motivations of the approach. In the previous section an approximation of the function $f(\underline{x})$ generating the data has been given. The argument of the approximating expression is a linear combination of the vector variable $\underline{x}$, which we call a ridge-type sum. The first generalization introduced in the next section involves the possibility of using hyperbolic functions, i.e., functions having products of variables as arguments. Then we define the class of functions which can be approximated by these methods. They are the multivariate functions of bounded variation. A function depending on many variables is called *multivariate*. The *bounded variation* is a property of such a function, which will be explained in the next section. If a function is of bounded variation then it is possible to substitute the usual differential dx in the integrals with $df(\underline{x})$, i.e., taking the small increments of $f(\underline{x})$ as a "differential." This concept is the basis of the theory of the Lebesgue–Stieltjes measure. The approximation of the function $f(\underline{x})$ is done by using a linear combination of increments of the function itself multiplied by the sigma function computed on products of variables. The main idea is as follows: suppose we know a function only in a finite set C of points of the domain K, then a linear combination of sigmoid functions evaluated on a hyperbolic function multiplied with the increments of the function among the set of points of C is used for approximating the function on every point of K. Such sums are best expressed in the form of a

Lebesgue–Stieltjes integral. This is done in Theorem 5.2 where the approximating formula to the function $f(\underline{x})$, known only in the points used to compute the increments $df(\underline{x})$, is written in the form of the Lebesgue–Stieltjes integral. In Theorem 5.2 such an integral is called the *approximation operator* and is denoted by the symbol $\Omega_\varrho(f)(x)$ where ϱ is a parameter. In Theorem 5.2 it is shown that this approximating form converges for each point of the domain K (i.e., pointwise convergence), while in Theorem 5.4 it is shown that the error made using this approximation can be made uniformly small. This estimate is *uniform* because it holds in any point of the domain where it is applied. To summarize, we can say that if we know the values of the function $f(\underline{x})$ only in some points of the domain K, then it is possible to construct an approximating formula for this function valid in all the points of the domain K using the approximation operator of Theorem 5.2 and knowing an upper bound of the approximation operator. If we compare the expression obtained in the formulation of the next section with the one of the previous section, we can say that the hyperbolic functions take the place of the ridge-type sums. The sigmoid functions are computed using as arguments the hyperbolic functions, while the synaptic weights v_k of the previous section are substituted by the increments of the known values of the function $f(\underline{x})$.

5.3 Approximation Operators

It is possible to use the sigmoid functions to generate approximation operators for multivariate functions of bounded variation. The idea for constructing a multivariate approximation procedure is to use a combination of one-dimensional functions together with appropriately chosen multivariate arguments. In this section we give the local and global approximation results for these operators, and we will see that it is possible to apply these results to the neural networks with one hidden layer and with fixed weights. The local approximation is discussed in Theorem 5.2 where the convergence of the approximation operator to the function $f(\underline{x})$ is shown. In Theorem 5.4 the maximum over all the points of K of the approximating error is analyzed: such an estimate can therefore be justly termed "global." The discussion about the approximation error for the approximation operator starts from Kolmogorov's mapping existence theorem for three-layer feed-forward neural networks ([24], [25]). As shown in the previous sections, the sigmoid functions are often used as NN input-output functions. Here the multivariate functions are not obtained by ridge-type sums as they are for the NN:

$$x \to \sum_{k=1}^{n} x_k v_k^{(i)}, \quad i = 1, 2, 3, \dots, \tag{5.3}$$

but by hyperbolic-type functions:

$$x \to \prod_{k=1}^{n} (x_k - v_k^{(i)}), \quad i = 1, 2, 3, \dots, \tag{5.4}$$

where $v^{(i)} \in R^n, i = 1, 2, 3, \ldots,$ are fixed vectors.

In order to prove the local and global approximation property of the combination of sigmoidal functions with multivariate arguments, we have to introduce some notation.

For $a, b \in R^n$ with $a \leq b$ $(a_i \leq b_i, 1 \leq i \leq n)$ we define:

$$[a, b] := \{x \in R^n : a_i \leq x_i \leq b_i, 1 \leq i \leq n\} \tag{5.5}$$

$$\mathrm{Cor}[a, b] := \{x \in R^n : x_i = a_i \vee x_i = b_i, 1 \leq i \leq n\} \tag{5.6}$$

$$\gamma(x, a) := |\{i \in \{1, \ldots, n\} : x_i = a_i\}|, \quad x \in \mathrm{Cor}[a, b], \tag{5.7}$$

where in (5.6) the symbol $\vee$ means the union of the one-point sets $x_i = a_i$ and $x_i = b_i$ and in (5.7) $|\cdot|$ indicates the number of different elements of the set under consideration. We can observe that $(n - \gamma(x, a))$ is the *Hamming distance* between x and a. We also indicate for a given function $f : R^n \to R^n$, the *interval function* Δ_f for all bounded intervals $[a, b] \subset R^n$ as follows:

$$\Delta_f[a, b] := \sum_{x \in \mathrm{Cor}[a,b]} (-1)^{\gamma(x,a)} f(x). \tag{5.8}$$

The symbol $[a, b]$ defines a multidimensional rectangle, $\mathrm{Cor}[a, b]$ the set of vertices of the edges of the rectangle and $\gamma(x, a)$ the "lower vertices" of the edges. Finally, $\Delta_f[a, b]$ is the sum of the differences of the values of the function $f(\underline{x})$ over the vertices of the rectangle $[a, b]$. So $\Delta_f[a, b]$ can be identified with the increments of the function mentioned in the previous section.

Definition 5.1 *A function $f : R^n \to R^n$ is said to be of (uniform)* bounded variation on R^n *($f \in BV(R^n)$) if there exists a constant $K \geq 0$ such that the interval function $\overline{\Delta}_f$ defined for all the intervals $[a, b] \subset R^n$ by*

$$\overline{\Delta}_f[a, b] := \sup \Bigg\{ \sum_{i=1}^{r} \left|\Delta_f[a^{(i)}, b^{(i)}]\right| : \\ \left([a^{(i)}, b^{(i)}] \subset [a, b],\ (a^{(i)}, b^{(i)}) \cap (a^{(j)}, b^{(j)}) = \emptyset,\ i \neq j\right), \\ 1 \leq i, j \leq r, r \in \mathbb{N} \Bigg\}$$

satisfies

$$\sup \left\{\overline{\Delta}_f[a, b] : [a, b] \subset R^n\right\} = K. \tag{5.9}$$

It is common knowledge that a function $f \in BV(R^n)$ induces a signed Borel measure m_f, called the *Lebesgue–Stieltjes measure*, associated with f, which determines the *Lebesgue–Stieltjes integral* with respect to f.

In order to have more general results and not be restricted to the special form of the sigmoid function used in Chapter 4, we introduce some generalized forms of the sigmoid function. We can choose for $\sigma(x)$ any continuous function dependent on a fixed parameter $r \in R$ such that

$$\lim_{\xi \to -\infty} \sigma_r(\xi) = 0 \quad \text{and} \quad \lim_{\xi \to +\infty} \sigma_r(\xi) = +1 \tag{5.10}$$

are considered. It is now possible to introduce the Lebesgue–Stieltjes convolution operator $\Omega_\varrho(f)$ induced by sigmoidal functions. We also call it the *Lenze operator* from the name of the person who introduced it ([36], [37]). We also use this term for the approximation operator defined in (5.15).

Theorem 5.2 *Let $\sigma : R \to R$ be a sigmoid function. For $f \in BV(R^n)$ with $\lim_{|t|\to\infty} f(t) = 0$, the operators Ω_ϱ, $\varrho > 0$,*

$$\Omega_\varrho(f)(x) := (-1)^n 2^{1-n} \int_{R^n} \sigma\left(\varrho \prod_{k=1}^{n} (t_k - x_k)\right) df(t), \ x \in R^n \tag{5.11}$$

are well-defined maps into the space of the continuous functions on R^n. They are also linear and, if $K \geq 0$ (5.9), they are bounded by

$$\sup_{x \in R^n} |\Omega_\varrho(f)(x)| \leq 2^{1-n} K \sup_{\xi \in R} |\sigma(\xi)|. \tag{5.12}$$

If f is continuous at the point $x \in R^n$, the following local approximation holds:

$$\lim_{\varrho \to \infty} \Omega_\varrho(f)(x) = f(x). \tag{5.13}$$

So we have to evaluate the convolution integral (5.11) in order to approximate numerically the function f. As a first step, we introduce the operators $\Omega_\varrho^{(h)}$, $\varrho, h > 0$:

$$\Omega_\varrho^{(h)}(f)(x) := (-1)^n 2^{1-n} \sum_{j \in \mathbb{Z}^n} \sigma\left(\varrho \prod_{k=1}^{n} \left(h\left(j_k + \frac{1}{2}\right) - x_k\right)\right) \Delta_f[hj, h(j+e)], \tag{5.14}$$

with $e := (1, 1, \ldots, 1) \in \mathbb{Z}^n$, $j = (j_1, \ldots, j_n)$. From this formula we can deduce the interpretation described in the previous section. First let us note that it is a finite sum approximating the integral appearing in the formula defining the Lenze operator and that the Lenze operator is the Lebesgue–Stieltjes integral generated by the function $f(\underline{x})$. From formula (5.14) it is clear that the increments of the function $f(\underline{x})$, $\Delta_f[hj, h(j+e)]$ play the same role of the synaptic weights of the formula (5.1), and that the sigmoid function is computed on the hyperbolic function $\varrho \prod_{k=1}^{n}(h(j_k + 1/2) - x_k)$. From formula (5.14) it follows that, if we know the function in the vertices of the segments $[hj, h(j+e)]$, we have an approximation for the value of the function $f(\underline{x})$ in any point of the domain K. The parameter ρ in the sums of (5.14) is used for making the limit $h \to 0$ for obtaining the Lebesgue–Stieltjes integral. However, the use of the approximation operator for predicting the values of $f(\underline{x})$ can be done only using the finite sums of formula (5.14); in fact, this is what we do in Chapter 8 for checking the results of NN application with this algorithm.

Defining also $\varrho := h^{-n}$, we obtain the operators $\Omega^{(h)}$ depending only on $h > 0$,

$$\Omega^{(h)}(f)(x) := (-1)^n 2^{1-n} \sum_{j\in\mathbb{Z}^n} \sigma\left(\prod_{k=1}^{n}\left(j_k + \frac{1}{2} - \frac{x_k}{h}\right)\right) \Delta_f[hj, h(j+e)]. \tag{5.15}$$

A bound of the type of (5.12) is valid for $\Omega^{(h)}(f)(x)$:

$$\sup_{x\in R^n} \left|\Omega^{(h)}(f)(x)\right| \le 2^{1-n} K \sup_{\xi\in R} |\sigma(\xi)|\,. \tag{5.16}$$

Introducing some additional notation ([35]) we can now see the local approximation properties of the operator $\Omega^{(h)}$, $h > 0$.

Let $x \in R^n$ be arbitrary and let

$$H(x) := \bigcup_{k=1}^{n} \{t \in R^n : t_k = x_k\} \tag{5.17}$$

be the *hyperstar* associated with x. We denote by

$$Q_+^0(x) := \left\{t \in R^n : \prod_{k=1}^{n}(t_k - x_k) > 0\right\} \tag{5.18}$$

$$Q_-^0(x) := \left\{t \in R^n : \prod_{k=1}^{n}(t_k - x_k) < 0\right\} \tag{5.19}$$

the positive (resp. negative) open set of quadrants of x. For each $x \in R^n$ it is possible to write the space R^n as a disjoint union of $H(x)$, $Q_+^0(x)$ and $Q_-^0(x)$:

$$R^n = H(x) \cup Q_+^0(x) \cup Q_-^0(x). \tag{5.20}$$

Let $BVC_0(R^n)$ be the space of the functions of $BV(R^n)$ which vanish at infinity,

$$BVC_0(R^n) := \left\{f \in BV(R^n) \bigcap C(R^n) : \lim_{|t|\to\infty} f(t) = 0\right\}. \tag{5.21}$$

Let m_f be the Lebesgue–Stieltjes measure induced by $f \in B$.

If $f \in BVC_0(R^n)$, $x \in R^n$, and $[a, b] \subset R^n$ we find that:

$$m_f([a, b)) = \Delta_f[a, b] = m_f([a, b]) \tag{5.22}$$

and

$$m_f(H(x)) = 0. \tag{5.23}$$

We then obtain the local approximation property of the operator $\Omega^{(h)}$, $h > 0$ given by (5.15).

Theorem 5.3 *Let $\sigma : R \to R$ be a sigmoidal function and $\Omega^{(h)}$, $h > 0$, be the family of operators defined in (5.15). Then for all $x \in R^n$ and $f \in BVC_0(R^n)$,*

$$\lim_{h \to 0_+} \Omega^{(h)}(f)(x) = f(x). \tag{5.24}$$

Usually it is important to find not only local results, but also global uniform approximation. To obtain these we define the Δ-continuity [53].

Definition 5.2 *A uniformly continuous function f is Δ-continuous on $[a, b]$ if for each $\epsilon > 0$ there exists a $\delta > 0$ such that for all the intervals $[c, d] \subset [a, b]$,*

$$\Delta[c, d] := \prod_{k=1}^{n} (d_k - c_k) < \delta \Longrightarrow \left|\Delta_f[c, d]\right| < \epsilon. \tag{5.25}$$

If $f \in BV(R^n)$ is continuous on $[a, b]$, the simple uniform continuity together with (5.8) implies the Δ-continuity of f on $[a, b]$ [35]. Let us also define the *supremum norm of f on R^n*:

$$\|f\|_\infty := \sup\{|f(x)| : x \in R^n\}. \tag{5.26}$$

These results make it possible to prove the following theorem for the global uniform approximation.

Theorem 5.4 *Let $\sigma : R \to R$ be a sigmoidal function and $\Omega^{(h)}$, $h > 0$ be the family of operators defined in (5.15). Then for all the functions $f \in BVC_0(R^n)$ the operators satisfy*

$$\lim_{h \to 0_+} \|f - \Omega^{(h)}(f)\|_\infty = 0. \tag{5.27}$$

In this section we have thus shown that the family of operators $\Omega^{(h)}$, $h > 0$ defined in (5.15) is a local and a global family of uniform approximation operators. In Chapter 8 we will consider these operators in the one-dimensional case and apply them to the SWH and SL time series.

6

Extreme-Value Theory

The statistics of waves is important in understanding the forces acting on the sea shore and for determining its evolution. Interaction among waves and winds is crucial for wave motion. Knowledge of the probability of occurrence of extreme events is necessary for designing secure structures in the sea environment. Extreme-value theory provides powerful tools to evaluate the probability of extreme events. In this chapter our aim is to collect several contributions to the theory of extreme events in order to make a self-contained exposition. We present a selection of the papers which seem best suited to our procedures, aims and tastes ([55], [56], [16], [19], [8], [29], [17]). Theorems are outlined without giving the proofs, which can be found in the quoted literature; we prefer to underline their importance for operations on the data. The chapter is divided in two parts. The first describes the method for deriving the distribution of the maxima in the case of independent random variables from the statistics of the exceedances of the time series over a certain threshold. This method is called POT (peak over threshold) and will be used in Chapter 9 to show the results for sea measurements. Section 6.1 also gives the fundamentals of the theory. The hypothesis of independent random variables is very restrictive and obliges the researcher to extract subsequences of i.i.d. variables from stationary processes, getting too few data in the case of long-range correlations. In the second part we deal with theorems and useful results for weakly dependent data. There is an introduction to each section with exact mathematical statements where the ideas are explained in simple and more intuitive terms.

6.1 The Case of i.i.d. Variables and the POT Method

6.1.1 Introduction

The aim of extreme-event analysis is to find the probability distribution of extreme events, which usually are also rare events. The knowledge of such a probability makes it possible to answer questions of the type: how probable will be the repetition of an extreme event after it has taken place? What is the average expectation time between two consecutive extreme events? In order to answer these questions one needs some quantitative probability estimate of the largest value associated to the phenomena of interest. The measure obtained by the buoys is the variable to be considered in the case of the SWH or SL. Since these measures are taken at different times, usually at regular time intervals, we have a sequence of random variables $\{X_i\}_{i=1}^n$ and the extreme event can be characterized by the maximum of such variables: $M_n = \max(X_1, \ldots, X_n)$. It is evident that the SWH or SL has a random character, it is enough to consider some of the graphs of one year measurements considered in Chapter 7. So the variable X_i is random, and we call it a random variable (r.v.). The r.v. X_i can be described by its probability distribution $F_i(x) = P(X_i \leq x)$ where $P(X_i \leq x)$ is the probability that the random variable X_i takes a value smaller or equal to x. Clearly, $F_i(x)$ is an increasing function of x which assumes the zero value if x equals the minimum value of X_i and the value 1 if x equals the maximum value of X_i. Note that we put a dependence on i in $F(x)$ because a priori we do not know if the SWH or SL values measured at different times have the same probability distribution. In order to apply the theory of extreme events we need $F_i(x)$ to be independent from the index i, i.e., all the r.vs. X_i have the same probability distribution. Random variables with same probability distribution are called identically distributed (i.d.). The other main information needed is concerns the probabilistic independence of the X_i. We cannot know a priori if the probability that the SWH X_1 is equal to 1 meter is independent of the SWH assumed by the sea 10 measures before or 1 measure before. Usually, in a sea state SWH are connected for some time. The probabilistic dependence among random variables is a fundamental property of the theory and is expressed in terms of their common probability distribution. Let I_i, I_j be two given intervals; the joint probability distribution of the random variables X_i, X_j is the probability $P(X_i \in I_i, X_j \in I_j)$. This is the probability that the two events $X_i \in I_i$ and $X_j \in I_j$ take place together. The independence of the two events can be expressed by means of the probability distributions:

$$P(X_i \in I_i, X_j \in I_j) = P(X_i \in I_i)P(X_j \in I_j). \tag{6.1}$$

In other words, we can say that the joint probability of two events is equal to the product of the probabilities of the single events. The equality $EX_iX_j = EX_iEX_j$ follows directly from this definition if the r.vs X_i have zero average. EX means the average of the r.v. X with respect to its probability distribution. The inverse is not true in general. One cannot derive the factorization of the probability of joint events in the product of the probabilities of the single events from the factorization of the averages. We call the random variables independent if their probability distributions satisfy the relation (6.1). We look for values of the SWH or SL X_i time series that are independent and identically distributed (i.i.d.). Thus we have taken measures which are far enough in time in the SWH or SL time series to obtain independent r.vs. This condition decreases the number of values available for analysis. So let us consider a time series of independent and identically distributed random variables (i.i.d.r.vs.) $\{X_i\}_{i=1}^n$ with common probability distribution $F(x)$. Let $M_n = \max(X_1, \ldots, X_n)$ be the maximum of these n random variables. Because of the independence, the probability distribution function of M_n is given by

$$P(M_n \leq x) = F^n(x).$$

In the book [34] many theorems and proofs can be found about the convergence of $P(M_n \leq x)$ to a definite limit. We will explain some of these theorems in the next section. In particular, we will show that the possible limit distributions $H_\xi(x)$ are only of three types which can be determined by different values of the parameter ξ. There are sets of distributions $F(x)$ converging to one of the three types, each of these sets being called a domain of attraction of the limit law $H_\xi(x)$. The most important problem in data analysis of extreme values is finding the exact limit distribution $H_\xi(x)$ directly from the values of the time series $\{X_i\}_{i=1}^n$. The POT method does this task. It is based on the observation, which is expressed in the form of a theorem in the next section, that the probability distribution of the events $X_i \geq u$, where u is a suitable threshold, uniquely determines the extreme distribution $H_\xi(x)$ (theorem of Pickand). The distribution of the events $X_i > u$ is called the "tail distribution" and specifically the generalized Pareto distribution (GPD). The free parameters characterizing the GPD can be found from the data by using the maximum likelihood method. The connection between the GPD and the extreme distribution $H_\xi(x)$ is found on the basis of the distribution of the points of the time series which are above a certain threshold. This is a Poisson distribution since the points are independent. The Poisson distribution arises naturally from studying, for example, the distribution of independent points in a segment: the number of points is Poisson distributed. These concepts are expressed in a more rigorous way in the next section.

6.1.2 The Results

This section begins with a basic and important result about the properties of the extreme events of i.i.d. random variables. This is the Fisher–Tippet theorem.

Theorem 6.1 *Suppose that there exist two sequences of real numbers $a_n, b_n > 0$ such that*

$$\lim_{n\to\infty} P\left(\frac{M_n - a_n}{b_n} \leq x\right) = \lim_{n\to\infty} F^n(a_n x + b_n) = H(x), \tag{6.2}$$

where $H(x)$ is a non-degenerate distribution function. Then $H(x)$ must be an extreme-value distribution.

If (6.2) holds, then F belongs to the maximum domain of attraction of H_ξ; so, we write: $F \in MDA(H)$. It is possible to choose a_n, b_n so that $H(x)$ can be described by means of the generalized extreme-value (GEV) distribution :

$$H_\xi(x) = \exp\left(-(1 + x\xi)^{-1/\xi}\right) \tag{6.3}$$

provided $(1 + \xi x) > 0$.

ξ is an important parameter characterizing the extreme distribution; it is called the shape parameter. The Weibull and Frechet types of distributions are given by

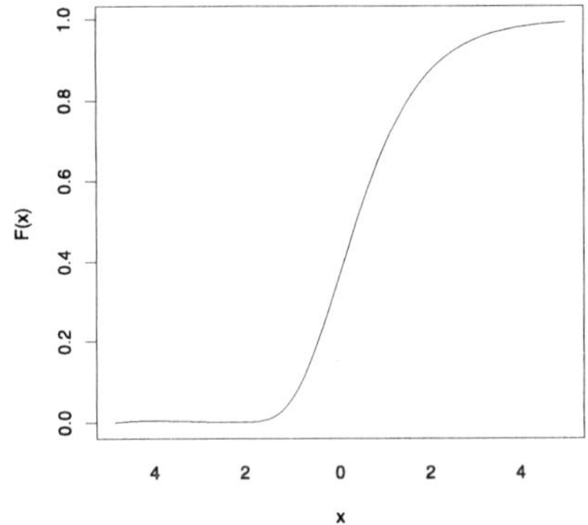

Fig. 6.1. Gumbel distribution function.

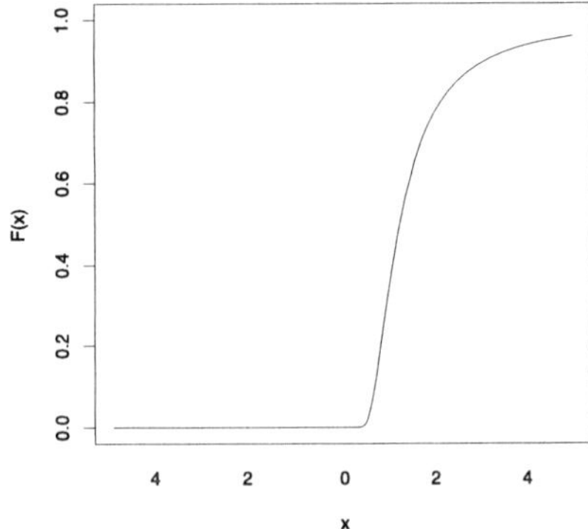

Fig. 6.2. Frechet distribution function.

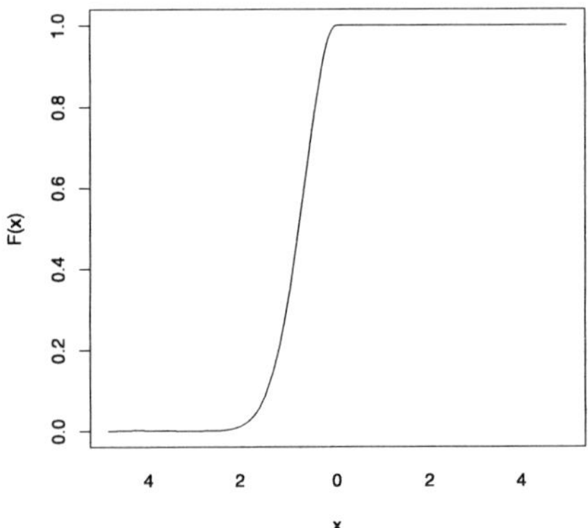

Fig. 6.3. Weibull distribution function.

(6.3) with $\xi < 0$ and $\xi > 0$, respectively. The Gumbel distribution type is obtained in the limit $\xi \longrightarrow 0$:

$$H_0(x) = \exp(-\exp(-x)). \tag{6.4}$$

In the Fig. 6.1 through 6.3 it is possible to see the graphs of the three distributions for particular values of ξ. The probability of extreme events is thus characterized by means of the GEV distributions. In particular, the inverse of ξ defines the behavior of the tail of these distributions. It is simple to see that the Weibull distribution has a finite right end point ($\omega = \{\sup\{x : F(x) < 1\}\}$) at $\omega = -1/\xi$. The Frechet distribution is heavy-tailed: in its domain of attraction there are distributions with tails that decay like a power function such as the Cauchy, Pareto and t-distributions. The Frechet and the Gumbel distribution have no finite right end point. In the domain of attraction of the Gumbel distribution there are distributions characterized by tails that decay faster than any negative power. Some examples are the gamma, normal, exponential and Weibull distributions. In the analysis of environmental time series, one must usually deal with sequences of values that are the realization of non-independent and non-identically distributed random variables. Let us suppose that a stationary time series can be obtained from the original ones, otherwise seasonal time series must be separately analyzed. After this, we can deal with the temporal dependence of the data. The effect of such dependence is the clustering of data, in particular around high values. One way to deal with clustered data is to decluster them properly. For this purpose, an early method comes from the blocks procedure proposed by Gumbel, within the theory developed for the extreme events of i.i.d.r.vs. Hereby, maxima from blocks of regular size are extracted. For environmental time series usually the size of the blocks is chosen to be a year, or the time-length of the characteristic seasons. Unfortunately, this procedure is not very efficient. The main drawback is that the extracted maxima could be a small data set; moreover, this declustering procedure does not ensure that the extracted values really correspond to independent events. In fact, with blocks of fixed size it is possible to select, for example, two maxima that actually belong to the same cluster of high values (for example, the SWH during a certain storm event). Clearly, a more efficient declustering procedure should be based on the feature of time series analyzed: for this purpose threshold methods are the best tools. The main idea of these procedures is to set a level that selects the values of interest.

The POT method is one of the main threshold methods, widely applied because it is a natural way to look at extreme events. In the early years the POT method was developed by hydrologists; subsequently, mathematicians worked to extend and show the proper way to define it. The procedure is simple: only the peaks of a group of exceedances over a given level should be selected, and only the ones separated by a sufficiently large time interval can be considered independent variables. The choice of the threshold and separation interval can be made considering the features of the phenomenon that generates the time series, and using the proper statistical diagnostic procedures.

Here the statistical model underlying the POT method is rigorously exposed. It will be applied in Chapter 9, to carry out the analysis of extreme events of sea storms at particularly significant Italian sites; diagnostic statistical tools will also be explained for the evaluation of the best-fitting model. The statistical model underlying the POT method is the Poisson-GPD model. A time series is a Poisson-GPD model if it has the following characteristics:

Definition 6.1 (The Poisson-GPD model)

(1) *Given a sequence (X_i), the exceedances over a certain threshold u are selected*

$$Y_{T_i} = X_{T_i} - u,$$

where $0 \le T_1 < T_2 < T_3 < \cdots$ are the exceedance times.

(2) *The exceedances over the threshold are i.i.d.r.vs.*

(3) *The sequence of interarrival times*

$$(T_1, T_2 - T_1, T_3 - T_2, \dots)$$

are independent r.vs. with common exponential distribution

$$F(x) = 1 - e^{-\lambda x}, \quad x \geq 0$$

with mean $1/\lambda$.

(4) *The number of observations occurring up to the time* t *constitutes a counting process*

$$N(t) = \sum_{i=1}^{\infty} I(T_i \leq t), \quad t \geq 0.$$

$N(t)$ *is a Poisson r.vs. with parameter* λt, *indeed the probability to have* k *observations up to the time* k *is given by*

$$P\{N(t) = k\} = \frac{(\lambda t)^k}{k!} e^{-\lambda t}.$$

(5) *Conditionally on* $N \geq 1$, *the exceedances over the threshold are i.i.d.r.vs. with a GPD.*

Environmental time series are not a realization of independent random variables. However, as already mentioned, by separating out the peaks within different clusters of exceedances, these will be approximately independent and therefore could fit a Poisson-GPD model. We now introduce the GPD. The functional form of this distribution is the following:

$$G_{\xi,\sigma}(y) = \left[1 - \left(1 + \xi \frac{y}{\sigma}\right)^{-1/\xi}\right], \quad y \geq 0, \ \sigma > 0,$$

where:

(a) if $\xi > 0$, then G is defined on $0 < y < \infty$, and the tail of the distribution satisfies

$$1 - G_{\xi,\sigma}(y) \simeq c y^{-1/\xi} \quad \text{with } c > 0,$$

that is, the usual "Pareto tail."

(b) if $\xi < 0$, the G has an upper end point at $\xi/|\sigma|$, and has a tail similar to the Weibull type of the traditional extreme-value distribution (EVD).

(c) if $\xi \to 0$, G converges to the exponential distribution

$$\left[1 - \exp\left(-\frac{y}{\sigma}\right)\right].$$

It is useful to observe that the mean of an r.v., Y, distributed following a GPD exists if $\xi < 1$, while the variance exists if $\xi < 1/2$. These quantities are given by

$$E(Y) = \frac{\sigma}{1-\xi}, \quad \text{var}(Y) = \frac{\sigma^2}{(1-\xi)^2(1-2\xi)}.$$

In the current literature one can see that for the SWH or SL time series usually $\xi < 1/2$.

At this point, we should show how this kind of distribution arises for exceedances over a high threshold. This is shown in Pickand's theorem.

Theorem 6.2 *Given a random variable X, let us introduce a threshold u and consider the variable $Y = X - u$. Let $F_u(y)$ be the conditional excess distribution of X, that is,*

$$F_u(y) = P\left[Y \le y \mid X > u\right], \quad y > 0.$$

Then $F \in MDA\left(H_\xi\right)$ if and only if there exists a function $s(u) > 0$ such that

$$\lim_{u \uparrow \omega} \sup_{0<y<\omega-u} \left|F_u(y) - G_{\xi,s(u)}(y)\right| = 0,$$

where ω is the right end point of $F(x)$.

Pickand's theorem can be considered as the basic motivation of the POT method. It suggests the following operations. Take the threshold u large enough or, more precisely, let it converge to ω, the right end point of the distribution $F(x)$, i.e., $u \to$ the first point where $F(x) = 1$, $F(x)$ is the probability distribution of the i.i.d. time series X_i. Consider the exceedances over the threshold u $Y_i = X_i - u$. If the exceedances are distributed in the limit as a GPD with certain parameters ξ, $s(u)$ (the σ parameter of the GPD is chosen as a function of the threshold), then the maximum of the time series X_i, $i = 1, \ldots, n$ has the distribution H_ξ for enough large values of n. Thus, in order to identify the distribution H_ξ of the extreme values of the time series X_i we need only to find the parameters ξ and $s(u)$ of the GPD distribution of the exceedances. The Poisson-GPD model can be fitted using the maximum likelihood method. In this case the likelihood function is written as a product of two different components, corresponding to the Poisson and the GPD distributions. If N exceedances $Y = (Y_1, Y_2, \ldots, Y_N)$ have been observed over a T-year period, then the likelihood function $\Phi_{N,Y}(\lambda, \sigma, \xi)$ is

$$\Phi_{N,Y}(\lambda, \sigma, \xi) = \frac{(\lambda T)^N e^{-\lambda T}}{N!} \prod_{i=1}^{N} g\left(Y_i; \sigma, \xi\right),$$

where $g\left(Y_i; \sigma, \xi\right)$ is the density of the GPD. Usually it is convenient to work with the log-likelihood function:

$$\begin{aligned} L_{N,Y}(\lambda, \sigma, \xi) &= \log \Phi_{N,Y}(\lambda, \sigma, \xi) \\ &= \log N \log \lambda + N \log T - \lambda T - N \log N - \log \sigma \\ &+ \left(1 + \frac{1}{\xi}\right) \sum_{i=1}^{N} \log \left(1 + \xi \frac{Y_i}{\sigma}\right) \end{aligned}$$

provided

$$\left(1 + \xi \frac{Y_i}{\sigma}\right) > 0$$

for all i.

The first term depends on λ and its maximum is given by

$$\hat{\lambda} = \frac{N}{T}$$

for which the variance, under the Poisson assumption, is

$$\operatorname{var}\left(\hat{\lambda}\right) = \hat{\lambda} T^{-2}.$$

Concerning the other terms, for $\xi < 1/2$ one can show that $\hat{\xi}_N$ and $\hat{\sigma}_N$, the best estimates of ξ and σ, have a neat asymptotic distribution

$$N^{1/2}\left(\hat{\xi}_N - \xi, \frac{\hat{\sigma}_N}{\sigma} - 1\right) \xrightarrow{d} N\left(0, \Sigma\right), \quad N \to \infty,$$

where the (2, 2) matrix Σ is

$$\Sigma = (1+\xi)\begin{pmatrix} 1+\xi & 1 \\ 1 & 2 \end{pmatrix}$$

and $N(\mu, \Sigma)$ is a bivariate normal distribution with mean vector μ and covariance matrix Σ. Again it is important to point out that for environmental time series $\xi < 1/2$ usually holds.

So after these fitting procedures, the parameters of the GPD are available and confidence intervals can also be evaluated. Let us note that the estimates of the parameters ξ and $\sigma = s(u)$ of the GPD identify the extreme distribution $H_\xi(x)$ of the maximum of the r.v. $X_1, \ldots, X_N$. The calculation of the best approximating values of ξ and σ is based only on the values of the measures and this is the basis of the functioning of the software R which directly gives the extreme distribution from the measures. These two parameters depend on the value of the threshold u because in the preceding formulas there are the r.vs. $Y_i = X_i - u$. The fact that the asymptotic distribution of the error of the estimates $\hat{\xi}_N - \xi$ and $\hat{\sigma}_N - \sigma$ is Gaussian with variance $N^{-1/2}$ ensures that asymptotically the estimates are good.

The Poisson-GPD model also makes it possible to establish the return values of extreme events and their confidence intervals as well.

Definition 6.2 *Let us define the m-year return level as the level expected to be exceeded in exactly one cluster of exceedances in m years.*

For example, ship structures are usually designed with respect to a 20-year return level. Unlike ship structures, offshore ones operate at fixed locations and are usually designed with respect to a 100-year return level. In general, these quantities are very important for engineering design activities: these are the values that engineers need to know from the statisticians.

Always within the POT framework, it is simple to observe that the expected number N_u of peaks over u in m years is

$$N_u = m\hat{\lambda}.$$

This formula can be understood from the estimate $\hat{\lambda} = N/T$ derived from the maximum likelihood method. In this formula, N is the number of exceedances over u in the period of T years. So one only has to use a proportionality argument to obtain this formula. With an analogous argument one finds that the expected number of observations exceeding x (with $x > u$) is

$$\left(m\hat{\lambda}\right)\left(1 - G_{\hat{\xi},\hat{\sigma}}(x)\right).$$

So, the m-years return level is obtained by setting this quantity equal to one and solving for x. Then, if $\hat{\xi} \neq 0$:

$$x_m^{(u)} = u + \frac{\hat{\sigma}}{\hat{\xi}}\left[\left(m\hat{\lambda}\right)^{\hat{\xi}} - 1\right],$$

otherwise:

$$x_m^{(u)} = u + \hat{\sigma}\log\left(m\hat{\lambda}\right).$$

The asymptotic variance of the return level estimates is

$$\text{var}\left(x_m^{(u)}\right) = d^T \Sigma d,$$

where d is the vector of derivatives of $x_m^{(u)}$ with respect to the estimated parameters ($\hat{\lambda}$, $\hat{\xi}$ and $\hat{\sigma}$ if $\hat{\xi} \neq 0$, $\hat{\lambda}$ and $\hat{\sigma}$ if $\hat{\xi} = 0$) and Σ is the total asymptotic (also including the Poisson component) covariance matrix, both evaluated by means of the estimates of the parameters.

6.2 Extreme-Value Theory for Stationary Random Processes

6.2.1 Introduction

In Section 6.1 we gave the definition of independent random variables which was necessary in order to apply the theorems and methods of Section 6.2. The definition is

$$P(X_i \in I_i, X_j \in I_j) = P(X_i \in I_i)P(X_j \in I_j). \tag{6.5}$$

But this condition cannot be verified for the usual SWH or SL if the distance among the indexes is not large enough. Thus, to apply the POT method we have to reduce the number of variables that we consider in order to obtain the independence. Consider also the fact that we have to study the variables above the threshold u and that u should tend to the extreme value of the distribution so that the number of events under consideration becomes small. One way to overcome this difficulty is to try to generalize the theorems and methods of the previous section to the case of dependent random variables. This problem is similar to the situation of the *central limit theorem (CLT)* where instead of the conditions (6.5) one can use the condition

$$|P(X_i \in I_i, X_j \in I_j) - P(X_i \in I_i)P(X_j \in I_j)| \to 0 \tag{6.6}$$

for $|i - j| \to \infty$ with a certain decay rate. But the theorems and methods of the extreme-value theory are not exactly the same as the corresponding arguments of the CLT. In fact, in this section we show that this condition, called weak mixing, is not sufficient for obtaining the convergence of the probability distribution of the maxima of the dependent r.v. to an extreme distribution. Thus, another condition is introduced in the next section: the anti-clustering condition. One of the main lemmas used for proving the convergence of the distribution of the maxima is the Poisson approximation. It connects the tail of the distribution $F(x)$ of the r.v. X_i with the probability distribution (PD) of the maximum of the r.v. $(X_1, \dots, X_n)$. The process consists of i.d. (identically distributed) r.v. so all the variables X_i have common PD $F(x)$. In the language of stochastic processes this property is equivalent to the stationarity. Let $\bar{F}(x) = 1 - F(x) = P(X \geq x)$ be the PD such that the random variable X is larger than x. Then the Poisson approximation is the property such that if $n\bar{F}(u_n) \to_{n\to\infty} \tau$ then $P(M_n \leq u_n) \to_{n\to\infty} e^{-\tau}$ and also the contrary holds. u_n is a sequence of thresholds. In other words this property means that if one studies the events over a certain sequence of thresholds and determines asymptotically the value τ of their PD, then the PD of the maximum of the n variables converges to the exponential of $-\tau$. In this way it is possible to find all the forms of the extreme distribution.

In the next section we show that the weak mixing condition is not strong enough for obtaining the extreme distribution of a stationary process, i.e., of a sequence of dependent r.vs., showing that the Poisson approximation does not hold in this case. A new hypothesis is thus introduced: the anti-clustering condition, such that if the process satisfies the weak mixing and anti-clustering condition then the extreme distributions for the maxima of the r.vs. of the process hold.

6.2.2 Exact Formulation

Let $X_1, X_2, \dots$ be a stationary process. We want to find conditions on the process such that the distribution function of the maximum $M_n = \max(X_1, \dots, X_n)$ converges to a Weibull or a Frechet or a Gumbel distribution function as in the case of i.i.d. random variables. The situation is somewhat similar to the CLT. Once it has been shown for a sequence of i.i.d. random variables, the theorem is extended to stationary processes if the random variables exhibit some properties such as the weak mixing condition. However, for the extreme-value theory it is necessary to introduce another hypothesis: the anti-clustering condition. We describe this approach in this section.

Condition 6.1 *Weak mixing condition $D(u_n)$.*

Let $p, q, i_1, \dots, i_p, j_1, \dots, j_q$ be any collection of integers such that

$$1 \le i_1 \le \cdots \le i_p \le j_1 \le \cdots \le j_q \le n$$

and $j_1 - i_p \ge l$. *Let*

$$A_1 = (i_1, \dots, i_p) \qquad A_2 = (j_1, \dots, j_q),$$

then the sequence $X_1, \dots, X_n$ *is weak mixing if*

$$\left| P\left(\max_{i \in A_1 \cup A_2} X_i \le u_n \right) - P\left(\max_{i \in A_1} X_i \le u_n \right) P\left(\max_{i \in A_2} X_i \le u_n \right) \right| \le \alpha_{n,l}$$

with $\alpha_{n,l} \to 0$ *for* $n \to \infty$ *for some sequences* $l = l_n = o(n)$.

Unfortunately, using the Poisson approximation, we show in the following example that this condition is not strong enough to obtain the required convergence. The Poisson approximation is the basic tool for the proofs of convergence of the distribution function of the maxima:

$$P(M_n \le u_n) \to_{n\to\infty} e^{-\tau} \quad \Leftrightarrow \quad n\bar{F}(u_n) \to_{n\to\infty} \tau$$

where u_n is a sequence of thresholds.

Example 6.1 *Let* Y_n *be i.i.d.r.v. with distribution function* $\sqrt{F}$. *Then the stationary process* $X_n = \max(Y_n, Y_{n+1})$ *satisfies the weak mixing condition but not the Poisson approximation.*

First note that X_n has the distribution function F.

$$\begin{aligned} P(X_n \le x) &= P(\max(Y_n, Y_{n+1}) \le x) = P(Y_n \le x, Y_{n+1} \le x) \\ &= P(Y_n \le x) P(Y_{n+1} \le x) = \sqrt{F(x)}\sqrt{F(x)} = F(x). \end{aligned}$$

Let us take $u_n : u_n \to x_F$ with x_F the right extremum of the distribution $x_F = \min(x \mid F(x) = 1)$. Thus, we have $F(u_n) \to_{n\to\infty} 1$. We choose the u_n in such a way that the condition

$$n\bar{F}(u_n) \to_{n\to\infty} \tau$$

holds for the variables $X_1, \dots, X_n$. Then we can show that the maximum M_n of the $X_1, \dots, X_n$ variables does not satisfy the Poisson approximation. We use the asymptotic relation

$$\begin{aligned} nP(Y_1 > u_n) &= n(1 - \sqrt{F(u_n)}) = n\frac{1 - F(u_n)}{1 + \sqrt{F(u_n)}} \\ &= n\frac{\bar{F}(u_n)}{1 + \sqrt{F(u_n)}} \to_{n\to\infty} \frac{\tau}{1+1} = \frac{\tau}{2}. \end{aligned}$$

We have to check that

$$n\bar{F}(u_n) \to \tau \not\Rightarrow P(M_n \le u_n) \to e^{-\tau}.$$

Let us take

$$P(M_n \le u_n) = P(\max(\max(Y_1, Y_2), \dots, \max(Y_n, Y_{n+1})) \le u_n)$$

remark: $\max(Y_1, Y_2), \max(Y_2, Y_3) = \max(Y_1, Y_2, Y_3)$

$$= P(\max(Y_1, \dots, Y_{n+1}) \le u_n) = P(\max(Y_1 \le u_n))^n \sqrt{F(u_n)}$$

$$= \sqrt{F(u_n)}^n \sqrt{F(u_n)} = e^{\frac{n}{2}\log(F(u_n))} = e^{\frac{n}{2}\log(1-\bar{F}(u_n))}$$

remark: $F(u_n) \to 1$ then $\bar{F}(u_n) \to 0$

$$\approx e^{-\frac{n\bar{F}(u_n)}{2}} \approx e^{-\frac{\tau}{2}}.$$

This argument suggests the introduction of another technical hypothesis.

Condition 6.2 *The anti-clustering condition $D'(u_n)$*

$$\lim_{k\to\infty} \limsup_{n\to\infty} n \sum_{j=2}^{[\frac{n}{k}]} P(X_1 > u_n, X_j > u_n) = 0.$$

Let us discuss the meaning of $D'(u_n)$. It implies that the exceedances of a pair of data above the sequence of thresholds become rare for large values of n:

$$E\left[\sum_{1\le i<j\le[\frac{n}{k}]} I_{\{X_i>u_n, X_j>u_n\}}\right] \le \left[\frac{n}{k}\right] \sum_{j=2}^{[\frac{n}{k}]} E[I_{\{X_1>u_n, X_j>u_n\}}] \to 0.$$

Remark 6.1 *$D'(u_n)$ does not hold in the example that we are discussing.*

Using the stationarity of the sequence we have

$$E\left[\sum_{1\le i<j\le[\frac{n}{k}]} I_{\{X_i>u_n, X_j>u_n\}}\right] = n \sum_{j=2}^{[\frac{n}{k}]} P(X_1 > u_n, X_j > u_n).$$

From the independence of the random variables X_1, X_j for $j > 2$ it is possible to write

$$n \sum_{j=2}^{[\frac{n}{k}]} P(X_1 > u_n, X_j > u_n)$$

$$= nP(X_1 > u_n, X_2 > u_n) + n\left(\left[\frac{n}{k}\right] - 2\right) P^2(X_1 > u_n)$$

$$= nP(\max(Y_1, Y_2) > u_n, \max(Y_2, Y_3) > u_n) + \frac{\tau^2}{2} + o(1)$$

$$\to_{k\to\infty,\ n\to\infty} nP(Y_1 > u_n) \to \frac{\tau}{2}$$

and hence the limit is not 0 and the condition $D'(u_n)$ is not satisfied.

For large values of n we can see that the pair of exceedances has the limit

$$E\left(\sum_{i=1}^{n} I_{\{X_i>u_n, X_{i+1}>u_n\}}\right) = nP(X_1 > u_n, X_2 > u_n)$$
$$= nP(\max(Y_1, Y_2 > u_n), \max(Y_2, Y_3 > u_n))$$
$$= \tau/2 > 0.$$

So, in the limit of long sequences the expected number of pairs of consecutive exceedances converges to $\tau/2$. We can now give the convergence theorem for the maxima of stationary processes. We use the symbol $MDA(H)$ to denote the maximum domain of attraction of an extreme distribution function H. We enunciate the main result, for the case of stationary sequences, about the asymptotic distribution of the maxima.

Theorem 6.3 *Let X_n be a stationary sequence with $F \in MDA(H)$ for any extreme distribution function H. Suppose that $\exists\ c_n,\ d_n$ such that*

$$\lim_{n\to\infty} n\bar{F}(c_n x + d_n) = -\ln(H(x)) = \tau \tag{6.7}$$

and that there exists $x \in \mathbb{R}$ such that $u_n = c_n + d_n x$ satisfies the conditions $D(u_n)$ e $D'(u_n)$. Then (6.7) is equivalent to

$$\lim_{n\to\infty} P\left(\frac{1}{c_n}(M_n - d_n) \le x\right) \to H(x).$$

Since the Poisson approximation holds for stationary processes satisfying $D(u_n)$ and $D'(u_n)$, the convergence of the sequence of maxima of such processes has the same limit as i.i.d. random variables. This theorem is analogous to the CLT for stationary processes in the sense that it is enough to check the mixing condition for the theorem to hold. In this case, however, we have an extra condition regarding the small probability of events of pairs of exceedances above a certain threshold.

The two conditions $D(u_n)$, $D'(u_n)$ are non-trivial and difficult to check, but there are two cases which commonly occur in the applications where these conditions are not so difficult to verify. One case is that of of Gaussian processes and the other is the ARIMA models (Chapter 8) which are frequently used for approximating stationary sequences. In these cases, it is possible to check the conditions $D(u_n)$, $D'(u_n)$ by inspection of the asymptotic behavior of the covariance:

$$\gamma(h) = \operatorname{cov}(X_0, X_h).$$

6.3 Process of Exceedances

6.3.1 Introduction

The condition $D'(u_n)$ concerns the pair of exceedances of the process above a certain threshold, i.e., the event of two variables having values larger than the threshold when the threshold grows for $n \to \infty$. This condition means that this event becomes rare when the threshold grows, something which we expect to hold for the usual time series of SWH or SL, although it is difficult to check directly. In this section we introduce the concept of point processes, which is useful for explaining the connection of the exceedances with the theory of extreme values. A point process is simply a configuration of points in the n-dimensional space taken randomly in a bounded set. The first important remark about the exceedances is that for i.i.d.r.vs. the exceedances are a point process distributed according to a Poisson process. In the following we want to make clear the equivalence of values above a threshold and point processes. We will use this connection to determine the distribution properties of the maxima of i.i.d.r.vs. The same can be done for stationary processes. The properties of exceedances seen as a point process are useful for extending the i.i.d. results to the case of dependent r.v. We start with the definition of point processes and with their convergence to a Poisson process. We then define the process of exceedances as a particular point process and discuss the theorems on the asymptotic distribution of maximum values for dependent r.v. using these concepts.

6.3.2 Point Processes and Poisson Processes

Point processes are generated by randomly casting points in a space R^n and counting how many of them belong to some finite number of given sets. The same description is used for a gas of atoms diffusing in a certain volume if one considers in which way the particles divide themselves among the subsets of the volume.

Clearly, the number of points in a set (finite or infinite) is a random variable and the study of the probability measure of these r.v. is the main object of this section.

Definition 6.3 (Counting Measure) *Given a sequence of r.v.* $\{X_i\}_{i=1}^n$ *in* $E \subset \mathbb{R}$ *let* $m(A)$ *be defined by*

$$m(A) = \sum_{i=1}^{n} \chi_{X_i}(A) \qquad \chi_{X_i}(A) = \begin{cases} 1 & \text{if } X_i \in A \\ 0 & \text{otherwise} \end{cases},$$

we have that $m(A)$ *is a counting measure,* $\forall$ *A Borel set of* $\mathbb{R}$*, and* $m(A)$ *is a point measure if* $m(K) < \infty, \forall$ *compact* $K \in E$.

We provide the following definition.

Definition 6.4 $M_p(E)$ *is the space of all point measures on E equipped with a given* σ*-algebra* $\mathcal{M}_p(E)$.

These definitions are the exact mathematical formulation of the following natural remarks.

Remark 6.2 $m(A) = N(A) = \sharp\{$*points belonging to the set* $A\}$.

Remark 6.3 *The* σ*-algebra* $\mathcal{M}_p(E)$ *is constituted of the sets* $\{m \in M_p(E) : m(A) \in B\} \forall B \subset [0, \infty]$.

Definition 6.5 (Point Process) *A point process N is a mapping from the probability space to the space of counting measures*

$$(\Omega, \mathcal{F}, P) \to [M_p(E), \mathcal{M}_p(E)].$$

Poisson process

We can now give the definition of a Poisson process as a particular case of a point process.

Definition 6.6 (Radon Measure) *μ is a Radon measure on $E \Leftrightarrow \forall$ compact $A \subset E$ $\mu(A) < \infty$.*

Definition 6.7 *A point process N is a Poisson process or a Poisson random measure with average Radon measure μ (PRM(μ)) if*

1. $\forall\ A \in E\ P(N(A) = k) = \begin{cases} e^{-\mu(A)} \frac{(\mu(a))^k}{k!} & \text{if } \mu(A) < \infty \\ 0 & \text{if } \mu(A) = \infty \end{cases}.$
2. *$\forall\ m > 1$, $A_1, \dots, A_m$ disjoint sets in E, $N(A_1), \dots, N(A_m)$ are independent r.v.*

Weak convergence of point processes

Let N_n be a sequence of point processes; the weak convergence of point processes to the point process N means that $\forall\ A_1, \dots, A_m$:

$$P(N_n(A_1), \dots, N_n(A_m)) \to P(N(A_1), \dots, N(A_m)).$$

Process of exceedances

We start the definition of this useful concept by using the point process of exceedances of an i.i.d.r.v. We introduce the usual symbols $(X)_n = X_1, X_2, ...$ i.i.d.r.v. with distribution function F and $M_1 = X_1$, $M_n = \max_{i=1,\dots,n}(X_i)$.

Definition 6.8 *Given a certain sequence of r.v. $X_1, \dots, X_n$ and a threshold u we define the point process $N_n()$ of the exceedances of $X_1, \dots, X_n$ above the threshold u as*

$$N_n() = \sum_{i=1}^{n} \chi_{\frac{i}{n}}() I_{\{X_i > u\}}, \quad n = 1, 2, \dots,$$

where ( ) is any subset of the interval (0, 1].

The r.v. $N_n(\ )$ takes integer values and is a counting measure in the sense that it counts the indexes of the sequence $1, \dots, n$ such that the corresponding variables $X_1, \dots, X_n$ are above the threshold u. The events defined by $N_n(0, 1] = k$ for different values of k define the interesting events for our analysis.

Example 6.2

$$\{N_n(0,1]=0\}=\left\{\sum_{i=1}^{n}\chi_{\frac{i}{n}}(0,1]I_{\{X_i>u\}}=0\right\}$$
$$=\left\{\sum_{i\le n}I_{X_i>u}=0\}=\{X_1\le u,\dots,X_n\le u\right\}=\{M_n\le u\}.$$

Thus, $\{N_n(0,1]=0\}$ is the event s.t. the maximum of $X_1,\dots,X_n$ is lower than the threshold u.

Example 6.3

$$\{N_n(0,1]<k\}=\left\{\sum_{i=1}^{n}\chi_{\frac{i}{n}}(0,1]I_{\{X_i>u\}}<k\right\}$$
$$=\{\sharp\{X_i\ge u,\ i\le n\}<k\}$$
$$=\{X_{k,n}\le u\},$$

where $X_{k,n}$ is the reduced k-statistics. Thus, the event $\{N_n(0,1]<k\}$ is nothing else than the case in which the kth maximum is less than the threshold.

The connection among these definitions is that a sequence of exceedance processes $\{N_n\}$ converges weakly to a homogeneous Poisson process in the case of i.i.d. and under conditions $D(u_n)$ and $D'(u_n)$ for stationary processes. However these last conditions can be checked with elegant tools in the case of exceedances. We start the series of theorems with the definition of a homogeneous Poisson process.

Definition 6.9 (Homogeneous Poisson Process) *A homogeneous Poisson process $N(t)$ is a Poisson process such that*

- $N(0)=0$
- *it has independent and stationary increments*
- $P(N(t)-N(s)=k)=[(\lambda(t-s))^k/k!]e^{-\lambda(t-s)}$.

With these definitions it is possible to apply the results on the weak convergence of an exceedance process to a homogeneous Poisson process and to give explicit criteria for the convergence of the distribution function of maxima to some extreme distribution function H.

6.3.3 The i.i.d. Case

We start the discussion of this case showing how the Poisson approximation can be reformulated in terms of exceedances and how it takes on an interesting meaning directly connected with the distribution of data. As usual, let u_n be a sequence of thresholds, $\{X_i\}_{i=1}^n$ i.i.d.r.v., $M_n=\max(X_1,\dots,X_n)$.

Proposition 6.1

$$P(M_n \le u_n) \to_{n\to\infty} e^{-\tau}$$

if and only if

$$n\bar{F}(u_n) = E(\sum_{i=1}^{n} I_{\{X_i>u_n\}}) \to \tau.$$

Remark 6.4 *This proposition makes it possible to take τ as the average number of exceedances above the given threshold u_n and thus it is possible to obtain the value of $\tau = \tau(u_n)$ directly from the inspection of the data.*

We now understand the importance of the theorem on *weak convergence of the point process of the exceedances in the case of i.i.d.r.v.:*

Theorem 6.4 *Let X_n be a sequence of i.i.d.r.v. with distribution function F. Let u_n be a sequence of thresholds such that Proposition 6.1 holds. Let*

$$N_n(\ ) = \sum_{i=1}^{n} \chi_{\frac{i}{n}}(\ ) I_{\{X_i>u_n\}}$$

be the point process of exceedances. Then $\{N_n\}$ weakly converges to the homogeneous Poisson process on $E = [0, 1]$ with intensity τ. In other words N is a PRM($\tau|\cdot|$) (where $|\cdot|$ is the Lebesgue measure on E).

In more explicit terms,

$$P(N_n(A_1) = k_1, \dots, N_n(A_m) = k_m) \to_{n\to\infty} e^{-\tau|A_1|}\frac{\tau|A_1|^{k_1}}{k_1!} \cdots e^{-\tau|A_m|}\frac{\tau|A_m|^{k_m}}{k_m!}.$$

6.3.4 The Stationary Case

Let us enunciate here all the useful facts and hypotheses for the statement of this theorem.

1. $D(u_n)$ and $D'(u_n)$ hold.
2. If u_n satisfies $n\bar{F}(u_n) \to \tau \in (0, \infty)$, then there are approximately τ variables above u_n in $X_1, \dots, X_n$.
3. Point process of exceedances:

$$N_n(\ ) = \sum_{i=1}^{n} \chi_{X_i}(\ ) I_{\{X_i>u_n\}}.$$

Thus, we can enunciate the theorem on *the weak convergence of the point process of the exceedances for the case of a stationary random process.*

Theorem 6.5 *Let (X_n) be a strictly stationary random process and let u_n be a sequence of thresholds such that Proposition 6.1 holds together with $D(u_n)$ and $D'(u_n)$. Then $N_n \to^d N$ in $M_p(E)$, where N is a PRM on $E = (0, 1]$ with intensity τ.*

The practical importance of this theorem is that the average number of random variables of a sequence $(X)_n$ with values above the threshold also defines a Poisson random measure in the stationary case for large n. In this way it gives an interesting criterion for checking the asymptotic distribution of the maxima in terms of the distribution of points above the thresholds. If the conditions $D(u_n)$ and $D'(u_n)$ are not known, this theorem can only be used heuristically in order to study the probability distribution of maxima of dependent variables. Thus, we are again faced with the problem of these two conditions. We discuss them in the following section.

6.4 Extremal Index

6.4.1 Introduction

We now discuss the deviations from independence in a stationary process and how these deviations are connected with possible deviations from the extreme distribution of the maxima. Again, it is useful to start from the Poisson approximation. According to the Poisson approximation, if $n\bar{F}(u_n) \to \tau$ then $P(M_n \leq u_n) = e^{-\tau}$. The deviation from the extremal distribution can be measured using a parameter θ (called the extremal index) which appears in this way: $P(M_n \leq u_n) = e^{-\theta\tau}$. The extremal index connects the dependence among the data with the PD of the maxima, in the sense that it is a parameter determinable from the time series. It measures in a simple way the deviation from the distribution function of the maxima for i.i.d.r.vs. In this section we give the exact definition of extremal index, explain its connection with the exceedances and give heuristic methods for finding the estimate of the value of the extremal index directly from the measures $X_i, i = 1, \ldots, n$.

6.4.2 Summary of the Theory

Before starting the discussion about the extremal index, it is worthwhile to summarize the main definitions and theorems for understanding these new concepts. Let $(X)_n$ be a stationary sequence with distribution function $F(x)$. Let $M_n = \max(X_1, \ldots, X_n)$, $F(x) = P(X_i \leq x)$ and $\tilde{M}_n = \max(\tilde{X}_1, \ldots, \tilde{X}_n)$ with $\{\tilde{X}_n\}$ a sequence of i.i.d.r.v. with the distribution $F(x)$ as $X_1, \ldots, X_n$, $P(\tilde{X}_i \leq x) = P(X_i \leq x)$.

1. Let us assume that $F(x)$ satisfies the limit

$$n\bar{F}(u_n) \to \tau \in (0, \infty) \tag{6.8}$$

   for some sequence of thresholds (u_n).
2. If we take a sequence of i.i.d.r.v. $\tilde{X}_n$ with the same distribution function F we know that

$$\lim_{n\to\infty} P(M_n \leq u_n) = e^{-\tau}. \tag{6.9}$$

3. Then from Theorem 6.4 we see that the point process of the exceedances of the i.i.d.r.v. $\tilde{X}_n$
$$N_n(\ ) = \sum_{i=1}^{n} \chi_{\frac{i}{n}}(\ ) I_{\{\tilde{X}_i > u_n\}}$$
weakly converges to a homogeneous Poisson process with intensity τ.
4. Suppose that $\{X_n\}$ satisfies the conditions $D(u_n)$, $D'(u_n)$. Then $\{X_n\}$ has the same asymptotic behavior of the sequence of i.i.d.r.v. $\{\tilde{X}_n\}$. In particular, (6.8) implies (6.9) (Poisson approximation)
$$\lim_{n\to\infty} n\bar{F}(u_n) \to \tau \ \Leftrightarrow \ \lim_{n\to\infty} P(M_n \le u_n) = e^{-\tau}$$
and $N_n \to N$ where N is a homogeneous Poisson process with intensity τ.
5. $D(u_n)$ and $D'(u_n)$ do not hold in any case. Recall Example 6.1. Let (Y_n) be i.i.d. with distribution function $\sqrt{F(x)}$. Then the stationary sequence $X_n = \max(Y_1, \dots, Y_n)$ has distribution function $F(x)$ and its maximum is $M_n = \max(Y_1, \dots, Y_n)$. In this case the Poisson approximation does not hold:
$$\lim_{n\to\infty} n\bar{F}(u_n) \to \tau \ \Leftrightarrow \ \lim_{n\to\infty} P(M_n \le u_n) = e^{-\frac{\tau}{2}}.$$

We can consider the factor $1/2$ appearing in the exponent $e^{-\tau/2}$ as a correction of the Poisson approximation due to the dependence among the data. These considerations suggest the definition of an extremal index.

Definition 6.10 (Extremal Index) *Let (X_n) be a stationary sequence, $\theta > 0$. If for any $\forall\, \tau > 0$ there exists a sequence (u_n) such that*
$$\lim_{n\to\infty} n\bar{F}(u_n) = \tau$$
$$\lim_{n\to\infty} P(M_n \le u_n) = e^{-\theta\tau},$$
then θ is an extremal index for (X_n).

It can be shown that the index $\theta \in (0, 1)$, for the sequence $X_n = \max(Y_n, Y_{n+1})$ $\theta = 1/2, \theta = 1$ for i.i.d.r.v. and for weakly dependent r.v.

In the case of linear processes $0 < \theta < 1$, if (X_n) has extreme index θ and if $(\tilde{X}_n)$ is a sequence of i.i.d.r.v. with the same distribution function, then
$$\lim_{n\to\infty} n\bar{F}(u_n) = \tau \Rightarrow \lim_{n\to\infty} P(\tilde{M}_n \le u_n) = e^{-\tau} \Rightarrow \lim_{n\to\infty} P(M_n \le u_n) = e^{-\theta\tau}.$$

If $u_n = c_n x + d_n, \tau = -\log(H(x))$ with
$$H(x) = \begin{cases} e^{-(-x^{\alpha})} \\ e^{-x^{-\alpha}} \\ e^{-e^{-x}} \end{cases},$$

then $\lim_{n\to\infty} P(M_n \le u_n) \to e^{\theta \log(H(x))} = H^{\theta}(x)$. From the discussion about the point process we know that τ has the meaning of the average density of the exceedances. We call the group of exceedances a cluster, and so we can say from this definition that, if the average density of clusters above u_n is $\tau(x) = -\log(H(x))$, then the distribution function of the maxima is $H^{\theta}(x)$. This suggestive proposition has the natural consequence of looking at a distribution function of maxima by studying the average density of clusters of exceedances above a certain threshold $u_n(x) = c_n x + d_n$. One then arrives at the result by evaluating the exponent $e^{-\theta\tau(x)}$. We are going to discuss operative methods for obtaining the value of θ directly from the data. The other important fact is that the deviation from the law of maxima of i.i.d.r.v. is measured by the difference $1 - \theta$.

Remark 6.5 *A negative value of θ does not make sense because the limit distributions would not be integrable.*

We can also show the following.

Proposition 6.2 *The index θ is less than one.*

Proof. We have the following inequality of probabilities:

$$P(M_n \le u_n) = 1 - P\left(\bigcup_{i=1}^{n}(X_i > u_n)\right) \ge 1 - n\bar{F}(u_n).$$

In the limit $n \to \infty$ one has

$$\begin{aligned} e^{-\theta\tau} &\ge 1 - \tau \\ 1 - \theta\tau &\ge 1 - \tau \\ \theta &< 1. \end{aligned}$$

■

Remark 6.6 *Not all the extreme distribution functions have an extreme index.*

Consider (X_n) with distribution function $F \in MDA(\Phi_\alpha)$, where $\Phi_\alpha(x) = e^{-x^{-\alpha}}$ is the Frechet distribution function. Let c_n be the normalization constants necessary to obtain the Frechet distribution function, $A > 0$ a r.v. independent from (X_n).

$$\begin{aligned} &P\left(\frac{1}{c_n}\max(AX_1, \dots, AX_n) \le x\right) \\ &\quad = P(c_n^{-1}M_n \le A^{-1}x) \\ &\quad = E(P(c_n^{-1}M_n \le A^{-1}x|A)) \to E(\exp\{-x^{-\alpha}A^{\alpha}\}), \end{aligned}$$

where E is the average with respect to the r.v. A. It is evident that the obtained distributions are not Frechet.

Example 6.4 *Consider a dam built on the shore for flood protection for the next 100 years with a safety margin of* 95%. *Suppose that* 99.9%, 99.95% *of the waves are smaller than* 10 *and* 11 *meters, respectively. Then in order to have a safety margin of* 95% *for the next 100 years, supposing that the annual maxima are i.i.d.r.vs., an* 11*-meter-high dam is sufficient because* $(0.9995)^{100} \simeq 0.95$. *But if the extreme index is* $\theta = 1/2$ *then a* 10*-meter-high dam is sufficient since* $(0.999)^{50} \simeq 0.95$.

6.4.3 Practical Estimates of θ

We describe two heuristic methods for the determination of θ. The first one starts from the definition of θ. If $n\bar{F}(u_n) \to \tau > 0$, then

$$P(M_n \le u_n) \simeq P^{\theta}(\tilde{M}_n \le u_n) = F^{\theta n}(u_n)$$

and so

$$\lim_{n\to\infty} \frac{\ln P(M_n \le u_n)}{n \ln F(u_n)} = \theta.$$

This limit suggests a simple procedure for the evaluation of θ starting from the data.

$F(u_n)$ and $P(M_n \le u_n)$ are to be estimated starting from the values of the time series. From the very definition of extremal index we obtained

$$P(M_n \le u_n) \simeq P^{\theta}(\tilde{M}_n \le u_n) = F^{\theta n}(u_n),$$

where $\tilde{M}_n = \max(\tilde{X}_1, \dots, \tilde{X}_n)$ are i.i.d. with the same distribution function F as X_i.

From the preceding discussion we know that this relation holds if

$$n\bar{F}(u_n) \to \tau,$$

where τ is given from the mean value of the number of data with values over the thresholds u_n

$$E\left(\frac{1}{n}\sum_i I_{\{X_i > u_n\}}\right) = \tau.$$

$\bar{F}(u_n)$ can be computed by

$$\bar{F}(u_n) = \frac{\tilde{n}}{n} = \frac{1}{n}\sum_{i=1}^{n} I_{\{X_i > u_n\}}.$$

The estimate of $P(M_n \le u_n)$ is not as easy. From the weak mixing condition $D(u_n)$ we find that

$$P(M_n \le u_n) \approx P^k(M_{[\frac{n}{k}]} \le u_n).$$

Let $n = rk, r = [n/k]$; then we can divide the whole sequence in k blocks of length r: $X_1, \dots, X_r, X_{r+1}, \dots, X_{2r}, X_{(k-1)r+1}, \dots, X_{kr}$ and compute the maximum in each block:

$$M_r^i = \max(X_{(i-1)r+1}, \dots, X_{ir}), \quad i = 1, \dots, k.$$

Then we have

$$P(M_n \le u_n) = P\left(\max_{1\le i\le k} M_r^i \le u_n\right) \approx P^k(M_r \le u_n)$$
$$\approx \left(\frac{1}{k}\sum_{i=1}^{k} I_{\{M_r^i \le u_n\}}\right)^k = \left(\frac{k - \sum_{i=1}^{k} I_{\{M_r^i \ge u_n\}}}{k}\right) = \left(1 - \frac{\tilde{k}}{k}\right)^k,$$

where $\tilde{k}$ is the number of blocks such that $M_r^i \ge u_n$.

Combining these estimates we obtain for θ:

$$\hat{\theta}_n^1 = \frac{k \ln\left(1 - \frac{\tilde{k}}{k}\right)}{n \ln\left(1 - \frac{\tilde{n}}{n}\right)}.$$

If $\tilde{k}/k$ and $\tilde{n}/n$ are much smaller than one we can expand the logarithms:

$$\hat{\theta}_n^2 = \frac{k}{n}\frac{\tilde{k}}{k}\frac{n}{\tilde{n}} = \frac{\tilde{k}}{\tilde{n}}.$$

The block method is connected with the concept of clusters. The event $M_r^i > u_n$ is a consequence of the fact that there must be at least one datum, i.e., a cluster, in the ith block with values larger than the threshold:

$$\{M_r^i > u_n\} = \bigcup_{j=1}^{r'} \{X_{(i-1)r+j} > u_n\}$$

with $r' \le r$. If this event occurs we say that a *cluster* occurred in the ith block.

The second method is based on the evaluation of the average dimension of the clusters. We start from a theorem which shows that under stronger conditions than $D(u_n)$ the point process of exceedances

$$N_n = \sum_{i=1}^{n} \chi_{\frac{i}{n}} I_{\{X_i > u_n\}}$$

converges weakly to a compound Poisson process

$$N(\ ) = \sum_{i=1}^{\infty} \xi_i \chi_{\Gamma_i}(\ )$$

if $n\bar{F}(u_n) \to \tau > 0$, where ξ_i represents the multiplicity of the cluster and Γ_i are the values of the variables belonging to the cluster which form a homogeneous Poisson process $N(\ )$ for large n.

The homogeneous Poisson process $N(\ )$ has intensity $\theta\tau$ and the number of points ξ_i in the ith cluster have the distribution function π_j[1] on N

$$E(N(0,1]) = E\left(\sum_{i=1}^{\infty} \xi_i \varepsilon_{\Gamma_i}(0,1]\right) = E\left(\sum_{i=1}^{\infty} E(\xi_1)\varepsilon_{\Gamma_i}(0,1]\right) = \theta\tau E(\xi_1),$$

where $E(\xi_1)$ is the average dimension of the clusters

$$E(\xi_1) = \sum_{j=1}^{\infty} j\pi_j.$$

From these results we easily establish that

$$\begin{aligned}\tau &= \lim_{n\to\infty} n\bar{F}(u_n) = \lim_{n\to\infty} E(N_n(0,1]) \\ &= \lim_{n\to\infty} E\left(\sum_{i=1}^{n} I_{\{X_i>u_n\}}\right) = E(N(0,1)) = \theta\tau E(\xi_1)\end{aligned}$$

so that $\theta E(\xi_1) = 1$. We get the nice interpretation that θ is the inverse of the average dimension of the clusters, so there is another estimator of θ:

$$\hat{\theta}_n^2 = \frac{\sum_{i=1}^{k} I_{\{M_r^i>u_n\}}}{\sum_{i=1}^{n} I_{\{X_i>u_n\}}} = \frac{K}{N}.$$

In other words, $\hat{\theta}_n^2$ is obtained by dividing K, the number of clusters which contain some exceedances, by the total number of exceedances.

1

$$\pi_j = \lim_{n\to\infty} \pi_j(n) = \lim_{n\to\infty} P\left(\sum_{i=1}^{r} I_{\{X_i>u_n\}} = j) \,\middle|\, \sum_{i=1}^{r} I_{\{X_i>u_n\}} > 0\right)$$

with $j \in \mathbb{N}$ where $P(A \mid B)$ is the conditional probability of event A with respect to event B.

7

Application of ANN to Sea Time Series

In this chapter we will show the application of the algorithms and methods explained in Chapter 4 to the time series of sea level (SL) and sea wave height (SWH) measurements. As specified in Chapter 2, SL is the height of the tide, and SWH is the significant wave height. The phenomenologies of the two time series are different and each has its own problems.

7.1 The SWH Time Series and Its Correlation Properties

We already mentioned that SWH information is contained in four time series: the wave heights, their direction, the peak period and the average period. We collected time series from eight stations distributed around the coast of Italy and its islands. Each time series contains data of the last ten years. The longest series is about 30 years. Data regarding SWH are gathered every three hours. As an example for this analysis, let us consider the time series of measurements gathered from Alghero, a port in Sardinia that has a complex phenomenology depending on currents from different areas of the Mediterranean Sea. We analyzed data for the period ranging from July 1, 1989 to December 31, 1998. According to the definitions given in Chapter 2, the four SWH time series are defined by:

- Significant sea wave heights H_s (centimeters);
- Peak periods T_p (seconds);
- Average periods T_m (seconds);
- Average incoming wave directions D_m (degrees).

H_s, T_m, T_p are estimated from the wave heights $Z(t)$ sampled every 0.25 second for a time interval of $T = 64$ seconds. This procedure gives a block of 256 values which are then transformed using fast Fourier transform. The correlation function of the heights is expressed by the formula

$$C_{zz}(f) = E\left|\frac{1}{T}\sum_{t=0}^{T} Z(t)\exp(2\pi i f t)\right|^2, \tag{7.1}$$

where E is the expectation with respect to the process generating the data (see Chapter 2) which is supposed to be stationary. By making the square of the modulus and an integration in time it is possible to see that the definition in (7.1) is the usual Fourier transform of the autocorrelation $C(\tau) = EZ(0)Z(\tau)$. The frequencies f vary in the interval (0, 1) according to the discrete Fourier transform but real measurements of f belong to the interval $(0.05, 0.635)$ Hz. This interval is divided into 117 bands of amplitude 0.005 Hz. Thus, the correlation function is a vector of 117 components for the first block of 256 measurements taken each 0.25 second. The 9 vectors of such type obtained during the time of 30 minutes are averaged. This averaged correlation function $C_{zz}(f)$ is used for defining the preceding quantities. Expanding the square and making the averages we get

$$C_{zz}(f) = \frac{1}{T} \sum_{\tau=-T}^{T} C(\tau) \exp(2\pi i f \tau).$$

If we define, as we did in Chapter 2,

$$q_i = \int_0^1 df f^i C_{zz}(f),$$

we get the definition of the SWH quantities:

$$\begin{aligned} H_s &= q_0, \\ T_m &= \frac{q_i}{q_0}, \\ T_p &= \frac{1}{f^*}, \end{aligned} \qquad . \quad (7.2)$$

where f^* is the value of the frequency on which $C_{zz}(f)$ gets the absolute maximum.

The lag found for these four time series is large and approximately around 60. However, 60 is a too large number of input neurons and a neural network (NN) with such an architecture is not very likely to work. Moreover, we have seen from our estimates that, after a lag of 60, the correlation function starts to increase again so the correlation might be even longer. Then we studied the time series obtained by making the first difference on the original time series: $Y(t) = X(t) - X(t-1)$. This operation implies the elimination of the slowly varying components. $Y(t)$ is often less correlated than the time series $X(t)$. In fact, for each of the four time series (H_s, T_p, T_m, D_m) the series $Y(t)$ has a correlation decaying to zero after $k = 10$, see Fig. 7.1 through 7.5. The behavior of the function $r(k)$ remains the same if we start the evaluation from different data of the time series and also if we take data from different years. This also happens for the data of other stations. Thus, we can conclude that our time series are stationary at the level of two point correlations. Also the correlation of the first differences has a stationary behavior and its decay length is the same for all the data.

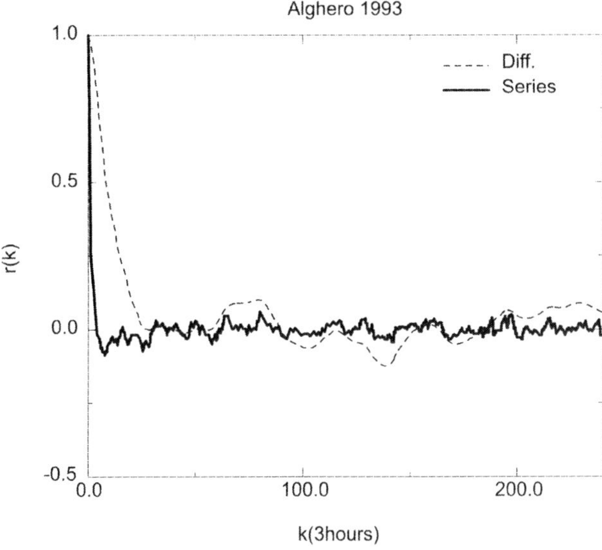

Fig. 7.1. Autocorrelation for the SWH of Alghero in January of 1993.

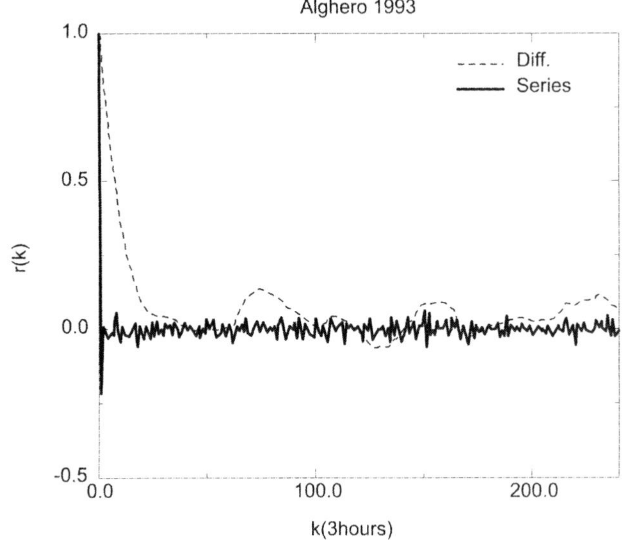

Fig. 7.2. Autocorrelation for the Peak Periods series T_p of Alghero in January of 1993.

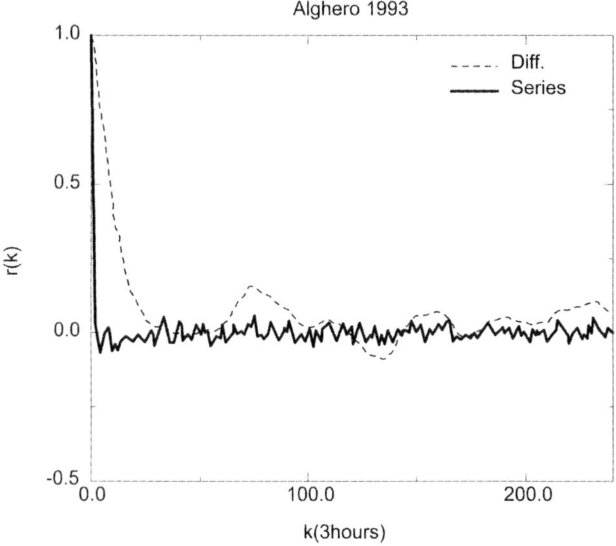

Fig. 7.3. Autocorrelation for the Average Peak series T_m of Alghero in January 1993.

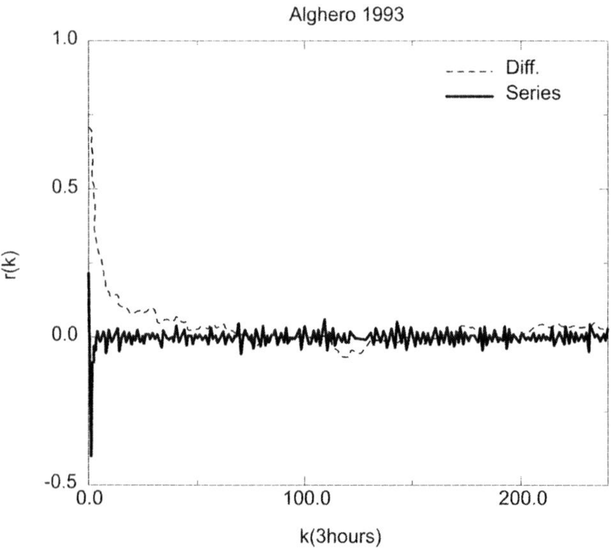

Fig. 7.4. Autocorrelation for the Average Directions of Alghero in January 1993.

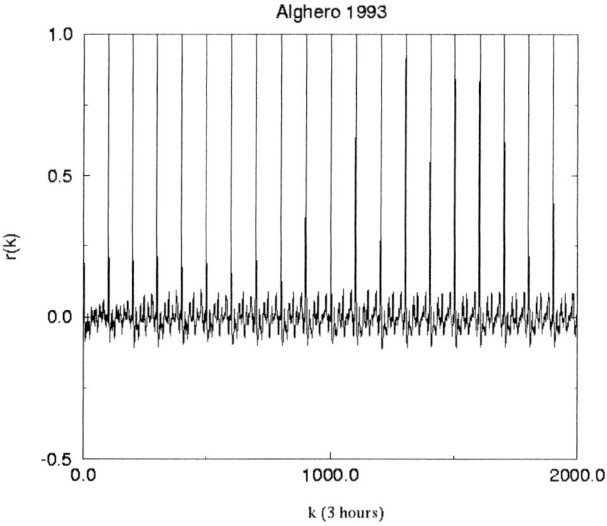

Fig. 7.5. Autocorrelation of the first difference of SWH for Alghero in January 1993.

7.2 The SL Time Series and Its Correlation Properties

The SL are dominated by astronomical periods and therefore have a strong periodic behavior. Meteorological conditions also have a strong influence although the main periods have an astronomical origin. The Porto Torres (a port in Sardinia) measurements give evidence of the non-stationary behavior of the correlations and the long decay length (Fig. 7.6). Thus, the embedding dimension is very large; moreover, in the case of periodic time series $z(t)$ the first difference $z(t) - z(t-1)$ still has a strong deterministic component and so for SL time series it is not useful to use this procedure.

7.2.1 The Input Vectors for SWH and SL

In both cases (SL and SWH) we used an NN for the reconstruction problem with inputs of data different from the time series; otherwise, the error of prediction increases strongly with the number of iterations. Every time new data are estimated they are affected by learning or training error. Thus, after n estimates, the error has become n times larger. Since in many cases the number of missing data is of the order of one thousand, in these cases such a reconstruction is flawed. To overcome this problem, we have used an NN which takes inputs from the time series of neighbhor stations and has, as output, the data for the station with missing data (see Section 7.3.1).

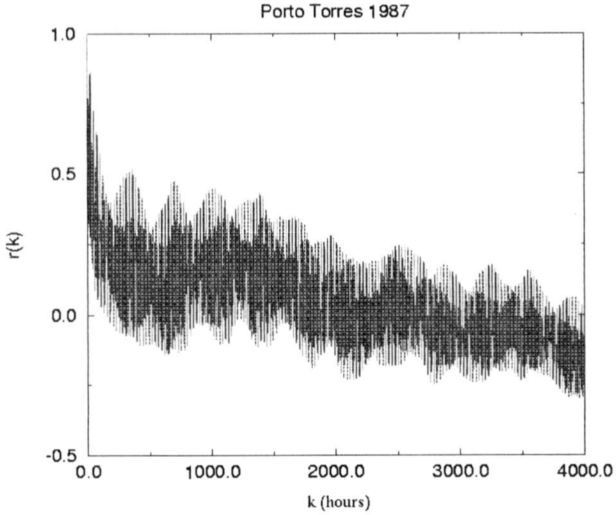

Fig. 7.6. Autocorrelation of SL for the data of 1987.

7.3 Neural Networks and Sea Time Series

In this section we describe the reconstruction of missing data in SL and SWH time series, using NNs. The Italian *Rete Mareografica Nazionale* (RMN) [National Sea Level Measurement Network] and the *Rete Ondametrica Nazionale* (RON) [National Sea Wave Measurement Network] were set up to obtain measurements of physical parameters crucial in defining the sea state. We analyzed in detail such methods in Chapter 2. Any possible failure and/or damage in measurement, transmission or data archiving instruments gives rise to gaps in the time series. The aim of the present work is to develop a suitable mathematical model in order to reconstruct missing data in time series, guaranteeing the necessary efficency. This model includes different and specific NNs. In the first subsection, the specific algorithm used to perform reconstruction of missing SL data will be described. The architecture of the used NNs, the advantages rising from the described approach and the results obtained will be included. In the second subsection, the specific algorithm used to perform reconstruction of missing SWH data will be provided, as well as the results of the ad hoc built adaptive NN ([77]). A system that allows the real-time reconstruction of missing data will be the topic of Section 7.4.

7.3.1 RMN and NN

The RMN measures physical parameters useful for defining the sea state. In this context the development of an NN system is very important for reconstructing missing data due to possible failure and/or damage of measurement instruments. The main aim is to estimate a suitable value for those data, by means of NNs, using as input the nearby stations' data. The input stations are chosen by taking into account geographical and statistical criteria.

More details about RMN and about the physical quantity of interest will be provided, as well as NN architecture and characteristics. The main focus will be on SL values and on the specific algorithm used to perform reconstruction of missing SL data during the period 1968–1999. The results obtained will be analyzed by means of statistical and mathematical tools and compared with those obtained by an astronomical model currently in operational service in Italy.

Measurement of the sea level

The RMN consists of 23 survey stations, uniformly distributed along the Italian coast. These stations are located in the Italian cities and locations cited in Section 2.6. Each station is provided with measurement, data control, transmission and storing instruments. The main physical quantities measured are: (i) air and water temperature, (ii) wind direction and intensity, (iii) atmospheric pressure and (iv) sea level. Each measurement is recorded every hour. Our focus is on sea level data, and our aim is the reconstruction of missing values of the time series. From inspection of the data we find two main components: the first, called "astronomical," is due to the moon gravitational influence, while the second, "meteorological," is strongly connected to the atmospheric pressure value. The astronomical component shows an evident periodicity rising from cycles of the moon, while the meteorological component shows a trend opposite to the pressure component.

The geographically close stations have almost equal components and show similar SL behavior. As shown later, this is of fundamental importance in our approach based on NNs.

The reconstruction approach

We reconstruct missing data in the SL time series by means of a two-layer backpropagation NN. As an output, the sea level value at the time t is obtained, using as input the sea level data of a nearby station at the times $t, t-1$ and $t-2$.

The architecture of this NN is composed of three linear neurons in the input layer, eight non-linear neurons in the hidden layer, and one linear neuron in the output layer. The best way to reconstruct missing data for a certain station is to use data from the nearby or more correlated stations as input of the NN. For such reasons these stations are called *reference stations*. This strategy guarantees the stability of the reconstruction error, since reconstructed data that cause error amplification in filling long time gaps are never used as input data. However, data reconstruction is not achievable if there are also missing data in the input time series; in order to reduce the effect of this limit, several reference stations, suitably chosen and ordered by correlation and geographical closeness criteria (Table 7.1) were used. An accurate definition of the dimension of learning and testing data sets is a critical issue because it is responsible for NN performance. It is achievable by running many preliminary

Table 7.1. Reference stations ordered according to priority (from left to right).

Station	Reference Stations
Ancona	Ravenna-Venezia
Bari	Vieste-Otranto-Taranto
Cagliari	Salerno-Palinuro-Porto Torres
Carloforte	Porto Torres- Palermo-Civitavecchia
Catania	Taranto
Civitavecchia	Napoli-Livorno-Palinuro
Crotone	Taranto-Catania-Otranto
Imperia	Livorno-Porto Torres
Livorno	Imperia-Civitavecchia
Messina	Catania
Napoli	Palinuro-Palermo-Salerno
Ortona	Vieste-Bari-Ancona
Otranto	Bari-Taranto-Catania
Palermo	Napoli-Palinuro-Salerno
Palinuro	Napoli-Palermo-Salerno
Porto Empedocle	Catania
Porto Torres	Imperia-Livorno
Ravenna	Venezia-Trieste-Ancona
Salerno	Palinuro-Napoli-Palermo
Taranto	Catania-Bari
Trieste	Venezia-Ravenna
Venezia	Trieste-Ravenna
Vieste	Bari-Ortona-Otranto

computer simulations and having specific knowledge about physical phenomena. By the end of our research, 1000 measurements had been taken as a learning set: this value is consistent with a time period of three “half lunations” (14 days each one), since it ensures a good learning process of the astronomical component. Because of the similar meteorological influence within groups of nearby stations, related information can be easily extracted through input data. Our testing data set consists of 1500 measurements following the learning data set. The minimization algorithm used during the learning phase is simulated annealing with the Geman and Geman schedule (Section 4.3.2). The error function is

$$E = \frac{1}{N} \sum_{i=1}^{N} |y_i - o_i| ,$$

where N is the number of patterns in the learning phase, y_i is the real datum and o_i is the NN reconstructed datum.

Results

The results of the NN approach applied to the RMN are shown in this section, described and analyzed by means of statistical and mathematical tools.

The missing data compared with the reconstruction data and mean learning and testing errors (averaged on missing data years) are described, respectively, in Tables 7.2 and 7.3. Time series have been reconstructed using for each station the reference stations' data (Table 7.1), which have been chosen by taking into account the quantity and the quality of reconstruction.

Table 7.2 allows us to assess the efficiency of the NN approach, which provided the reconstruction of more than 70% of missing data in most cases.

The data in Table 7.3 have been computed as the average of learning and testing errors estimated for each reconstructed year on each station. Since the data from the same station are not always available, the best and the worst results were not taken

Table 7.2. Missing data and reconstructed data.

Station	Missing Data (%)	Reconstructed Data (%)
Ancona	45.5	83.8
Bari	48.5	74.9
Cagliari	29.0	62.6
Carloforte	54.1	41.5
Catania	23.8	83.4
Civitavecchia	50.9	78.6
Crotone	38.9	79.2
Imperia	31.5	71.8
Livorno	33.5	90.6
Messina	34.9	67.2
Napoli	24.9	64.8
Ortona	53.8	88.5
Otranto	39.3	57.7
Palermo	67.2	70.7
Palinuro	37.1	74.9
Porto Empedocle	41.6	35.8
Porto Torres	61.7	65.3
Ravenna	45.6	84.7
Salerno	35.1	63.4
Taranto	38.3	57.2
Trieste	40.6	77.5
Venezia	45.3	86.3
Vieste	51.4	77.2

Table 7.3. Average absolute learning and testing errors.

Station	E_L (mm)	E_T (mm)
Ancona	59	76
Bari	49	56
Cagliari	39	61
Carloforte	30	35
Catania	35	44
Civitavecchia	38	48
Crotone	30	37
Imperia	37	40
Livorno	43	44
Messina	36	57
Napoli	32	44
Ortona	63	70
Otranto	48	45
Palermo	39	54
Palinuro	21	37
Porto Empedocle	37	40
Porto Torres	36	37
Ravenna	37	50
Salerno	30	42
Taranto	31	42
Trieste	51	67
Venezia	42	59
Vieste	40	54

into account in the evaluation of the errors. Indeed, in order to fill in as many as possible missing data, several different stations were used, choosing the best available one each time.

The analysis of these results has shown that the learning and testing errors fluctuate, respectively, around the mean values of 3.9 and 5.0 cm.

Fig. 7.7 and 7.8 show the NN reconstruction of sea level measure, compared to the astronomical model, using data from the Palermo and Otranto stations during the year 1999 as an example. The astronomical model reproduces the periodical component rising from moon cycles, but it does not succeed in reproducing the meteorological one. On the other hand, the NNs are able to reconstruct the entire behavior of the sea level, while also detecting its sudden changes.

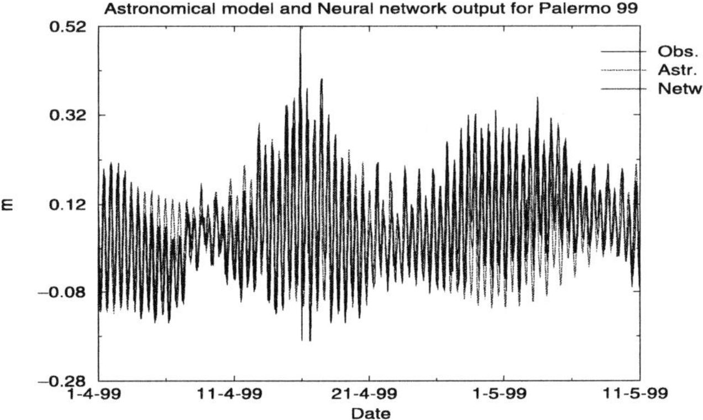

Fig. 7.7. NN reconstructed data vs. astronomical data, Palermo 1999.

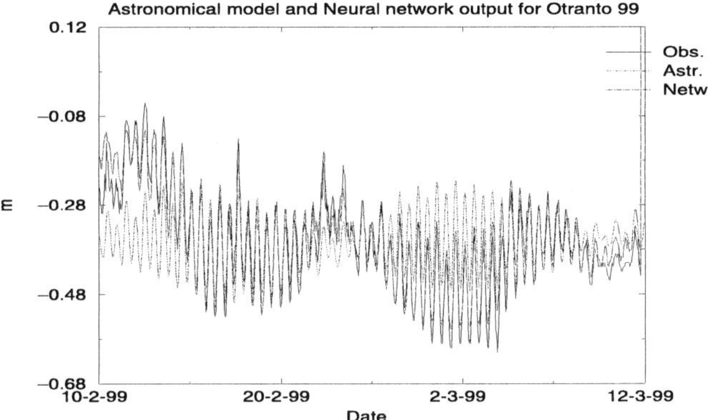

Fig. 7.8. NN reconstructed data vs. astronomical data; Otranto 1999.

Conclusions

An interesting result obtained in the context of this research is the error stability in the reconstruction of missing data in SL series.

As mentioned above, the NN system is able to prevent error amplification by using only real data (not reconstructed data), obtained through direct measurement provided by the nearest stations. Hence, an efficient reconstruction is also achieved for long missing data periods with a maximum length of six months. This limit is due to the length of the learning and training set. After a six-month reconstruction, it is necessary retrain the NN in order to obtain the highest efficiency.

Another remarkable result is the small error (Table 7.3), which has been estimated by averaging the errors on the reconstructed values using inputs from different reference stations belonging to the same group of reference stations. We note that the order of magnitude of the errors is the same as the one obtained by applying the expectation-maximization (EM) method (see Munk, Cartwright, and Ateljevich ([51], [52]) but our reconstruction has been done over much longer periods.

7.3.2 RON and NN

RON is designed to measure physical parameters in order to define the sea state; like RMN, damage and failures of measurement sometimes cause gaps in the time series.

Similarly to the case described above, a comprehensive NN system might fill gaps in time series with reconstructed data, ensuring a proper level of efficiency.

The significant wave height (SWH) is the parameter addressed in this research.

Measurements of the SWH

RON consists of several buoys uniformly distributed along the Italian coast. The aim of the research described in this paragraph is to reconstruct missing data in the SWH time series by means of NNs using real data measured by the buoys.

Most of the works reported in the literature propose a number of stochastic models, which proved to be useful in defining the probabilistic nature of the time series, but did not succeed in reconstructing missing data accurately. See, for example, the paper by Athanassoulis and Stefanakos ([9]).

On the other hand, numerical models are used to integrate wave motion equations, provide time localized, but not spatially resolved, information. This work provides a specific NN which is successful in reconstruction of wave heights, for each gap and for each buoy of the RON. The developed NN is called an *adaptive NN*, a name derived from the specific algorithm ([77]). This algorithm provides the relationship between nearby buoys' time series. The time-series analysis showed similar features and behaviors of time series referred to nearby buoys. For example, peaks in the Monopoli time series often correspond to peaks in the nearby Ortona and Crotone time series, and the same occurs in the calm periods of the wave motion. Hence, Crotone and Ortona have always been used as NN inputs for the Monopoli time series, becoming its reference bouys. Since the sea wave state is also defined by means of the incoming wave direction (D_m), we need to distinguish different waves by different incoming sectors. Thus, not only the significant wave height (H_s) but also the D_m have been considered as input variables. If the wave perturbation strikes different stations from different sectors, the NN determines how to select the wave height of the reference stations, reproducing the relationship among different zones. The best way to reconstruct missing data for a specific station is to use data from nearby or more correlated stations as input data ([49], [79]).

Results for the adaptive neural network

The definition of input variables is essential in determining the NN architecture. In this context, after preliminary trials, the input variables have been settled as the H_s and the D_m of two reference stations, at times $t, t-1$. The reference stations have been chosen on the basis of the similarity of the behavior of the SWH time series and of the physical phenomena. The output was the H_s at time t of the examined station. Another important task is to accurately define the dimension of the learning and testing data sets, as they are responsible for NN performance during the two phases described above. A comprehensive analysis of missing data periods for each station shows a constant number of gaps for each year, proving a stationary efficiency of the buoys.

The learning set consisted of 3000 measurements, corresponding to the first measurement year (we are referring to time series with data stored every three hours, in the period July 1,1989–December 31, 2000). The testing data set is taken as all the data following the learning data set. Based on preliminary tests, this result provides a good generalization error (GE) estimation.

The minimization algorithm used during the learning phase is simulated annealing with the Geman and Geman schedule. The error function that has to be minimized is

$$LE = \frac{1}{N} \sum_{t=1}^{N} p\,(H(t)) \cdot |H(t) - O(t)|\,,$$

where N is the number of patterns in the learning phase, H is the real datum H_s, O is the NN reconstructed H_s and $p(H)$ is a weight factor depending on the values of H.

The introduction of these weights has been crucial since it has drastically improved the NN performance, on the lowest as well as on the highest values of the SWH time series.

Different wave height intervals have showed a different number of occurrences: for instance, in Monopoli station, waves of 1 m take place 77% of the time. However, since high waves represent extreme events, the less probable they are, the more they have to be taken into account. The right compromise between the number of reconstructions and different weights for different events has to be reached.

The optimal choice of the coefficients $p(H_s)$ is:

$$p(H_s) = \begin{cases} P1 & \text{if } H_s \text{ belongs to the interval } [0, 200] \text{ cm} \\ P2 & \text{if } H_s \text{ belongs to the interval } [200, 400] \text{ cm} \\ P3 & \text{if } H_s \text{ belongs to the interval } [400, 600] \text{ cm} \\ P4 & \text{if } H_s \text{ belongs to the interval } [600, 800] \text{ cm,} \end{cases}$$

where $P1$, $P2$, $P3$ and $P4$ are defined after a number of learning attempts. For all the stations these coefficients are 2, 4, 16, 32, respectively.

After the minimization of LE, the mean absolute error has been evaluated (Table 7.4):

$$Em = \frac{1}{N} \sum_{t=1}^{N} |H(t) - O(t)|\,. \tag{7.3}$$

This error allows a direct comparison between the results obtained for each station, in the learning and in the testing phase. We conclude that a learning data set of just one year's measuring is enough to use the same NN with the same internal parameters for a maximum of nine consecutive years.

Table 7.4. Mean absolute error in learning and testing phase for one year testing for the period July 1, 1989–December 31, 2000.

Buoy	Me Learning (cm)	Me Testing (cm)
Alghero	52.1	53.7
Catania	28.0	30.4
Crotone	30.2	32.1
La Spezia	39.5	40.8
Mazara	38.7	39.5
Monopoli	26.3	29.1
Ortona	38.8	39.5
Ponza	29.9	32.3

The highest mean error is found in Alghero, the lowest in Monopoli. Similarly, Catania, Crotone and Ponza show an acceptable mean error. Table 7.5 shows the percentage of missing data for each station compared to the reconstructed ones.

In the case of Monopoli, a more accurate investigation has been completed, and other kinds of errors have been introduced: the mean signed error,

$$Es = \frac{1}{N}\sum_{t=1}^{N}(H(t) - O(t)), \tag{7.4}$$

and the mean relative error,

$$Er = \frac{1}{N}\sum_{t=1}^{N}\left(\left(\frac{|H(t) - O(t)|}{H(t)}\right) \cdot 100\right). \tag{7.5}$$

Table 7.5. Missing and reconstructed data percentages in the period July 1, 1989–December 31, 2000.

Buoy	Missing Data (%)	Reconstructed Data (%)
Alghero	3.7	79.9
Catania	10.2	62.9
Crotone	6.3	71.9
La Spezia	7.1	84.7
Mazara	16.4	74.5
Monopoli	5.7	81.2
Ortona	7.2	92.7
Ponza	9.3	76.1

Table 7.6. Adaptive NN on Monopoli: learning for the period 1990–1995. Correlation between real data and reconstructed data = 0.75.

Wave Height (cm)	Number of Data	Em (cm)	Es (cm)	Er (%)
<100	10,523	16	−3	48
100–200	2529	41	22	32
200–300	338	72	41	32
300–400	63	96	69	29
400–500	4	125	125	30
All	13,457	23	3	44

The correlation between real and reconstructed data Table 7.6 and Table 7.7 show the results obtained extending the learning phase to the period from 1990 to 1995 and the testing phase in the period 1996–2000.

The NN has been trained to improve reconstructions for highest values rather than lowest values.

The correlation values between real and reconstructed data are higher than 75% and the relative error is not higher than 32% and not less than 20% for waves higher than 2 m. Fig. 7.9 shows the results reached by the NN in comparison with the ones obtained by the numerical wave amplitude model (WAM); these results were obtained during the testing phase (learning of 3000 measurements, from July 1, 1989) in the period December 29, 1994–February 7, 1995. During this period an extreme event of 5 m wave height occurred. This comparison has been made in order to check the orders of magnitude of the values obtained from the NN reconstruction. Since the WAM predicts values of SWH, it can also be used for reconstructing the values. It is sufficient to consider the reconstructed values as predictions made by the WAM on the values computed before the gap. This is only a check of the order of magnitudes because the lattice point of the WAM does not coincide with that of the buoys.

Table 7.7. Adaptive NN on Monopoli: testing for the period 1996–2000. Correlation between real data and reconstructed data = 0.77.

Wave Height (cm)	Number of Data	Em (cm)	Es (cm)	Er (%)
<100	9467	17	−3	47
100–200	2124	41	26	31
200–300	384	72	41	31
300–400	71	93	77	29
400–500	6	83	77	20
All	12,052	23	4	44

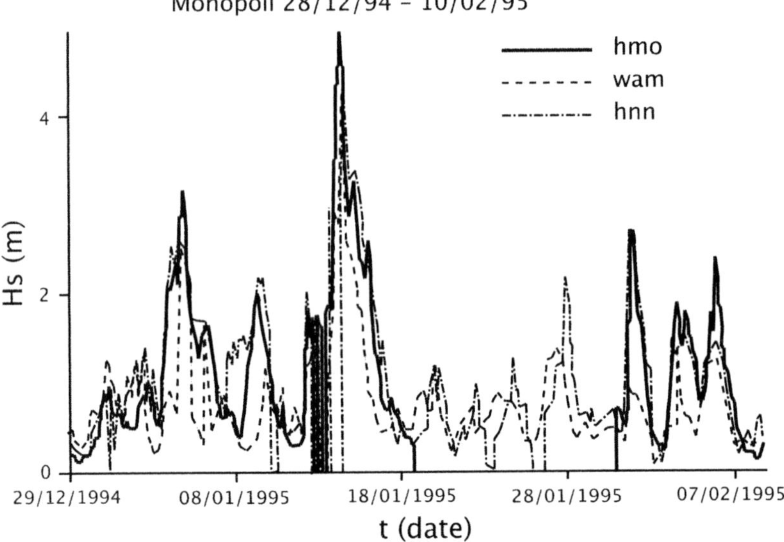

Fig. 7.9. hm0 represents the wave height referred to Monopoli buoy (real datum); hnn represents the NN reconstructed datum; wam represents the WAM reconstructed datum.

Results for the seasonal approach

A different application of the adaptive NN with specific features in each different season is found in the *seasonal adaptive neural network*. A new training procedure has to be defined in order to optimize the NN performance in each season. Indeed, different trends correspond to different seasons.

The time series (real data in comparison with the monthly averages) graphical analysis, is referred to the three seasons identified in 1998:

1) I season (IS): January, February, March and April
2) II season (IIS): May, June, July and August
3) III season (IIIS): September, October, November and December

Table 7.8 shows the learning and testing periods referred to each season; Table 7.9 shows the results obtained in Crotone and Ortona.

In this case, the error function coefficients in the learning phase are:

$$p(H_s) = \begin{cases} N/N1, & \text{if } H_s \text{ belongs to the interval } [0, 200] \text{ cm} \\ N/N2, & \text{if } H_s \text{ belongs to the interval } [200, 400] \text{ cm} \\ N/N3, & \text{if } H_s \text{ belongs to the interval } [400, 600] \text{ cm} \\ N/N4, & \text{if } H_s \text{ belongs to the interval } [600, 800] \text{ cm}, \end{cases}$$

where N is the number of patterns in the learning phase, while $N1$, $N2$, $N3$, $N4$ are the numbers of H_s belonging to each different wave height interval.

Table 7.8. Learning and testing periods for the seasonal NN.

Season	Learning	Testing
IS	January	April
	February	
	March	
IIS	May	July
	June	August
	September	
IIIS	October	December
	November	

Table 7.9. Learning and testing mean absolute errors for the seasonal NN in 1998.

Buoy	ME Learning (cm)	ME Testing (cm)
Crotone		
IS	23	21
IIS	20	21
IIIS	21	29
Ortona		
IS	41	25
IIS	18	20
IIIS	27	41

The best results were obtained in two different cases, Crotone and Ortona, using the seasonal NN (Table 7.9) in comparison with the results obtained with the adaptive NN (Table 7.4). Hence, the more specific the NN, the better the results obtained.

This research suggests that we distinguish different seasons in past years, train each NN in a specific way for each season and use each NN for the corresponding data reconstruction. The analysis described here is just the beginning of a topic which will be further developed in Chapter 9.

Conclusions

Our aim was to reconstruct the missing data at each time t and at each site where they occurred, in order to achieve a high spatial and temporal resolution.

The first significant result is that in many cases (Catania, Crotone, Monopoli and Ponza) we obtained correct reconstructions. Second, and even more meaningful, is that our approach seems to guarantee long-term efficiency. The stability of the error allows us to train the NN on a period of one year for the learning phase, and reconstruct data for the following years, without increasing uncertainty. A final advantage of this methodology is the possibility to include it in a computer system using NNs in real-time data reconstruction.

7.4 Real-Time Application of Artificial Neural Networks

In this section, we explain briefly the design of a software application that uses the artificial neural networks (ANNs) for a reconstruction of the sea data in a real-time measurement system. We shall focus on the high-level design of the flow of logic which has been devised for the Italian Sea Measurements Systems.

The same principles of operation are valid for similar real-time measurement systems. Usually, the measurements executed by such systems are stored in a central database, which is regularly updated. From this database the inputs to the NNs can be extracted. Thus, the corresponding neural outputs—in our case the reconstructions of SL and SWH—can be written into a dedicated database for further analysis. In order to provide a continuous operation, the action of the NNs has to be synchronized with the updating of the central database. In addition to this, some recovering procedures and proper messages should be issued to the user.

Once a training phase has been executed, the application of an NN is straightforward. Given the input data, the corresponding output can be computed by means of simple and fast computing steps. This is valid for the algorithms we have introduced in the previous sections. In our case, we used a standard PC (PENTIUM processor). A reconstruction of a single datum (of the SL or of the SWH) is obtained in less than a second, while updating the central database of the Italian Sea Measurement Systems can be completed in half an hour. This means that the NNs have no time problems in providing their outcomes.

Having said that, we distinguish three main tasks in this application; a module of specific procedures can be dedicated to each task. A first module responsible for reading new input data into the neural networks, a second module for executing the proper neural networks, and a third module for writing new reconstructions into a dedicated database. Some coordination among these modules is also required. For example, if for any reason it is impossible to write the reconstructions of some data, this should not block the reading of a new group of inputs and the subsequent activation of the neural algorithms. On the other hand, if it is impossible to read a new input, this should not block the writing of the previous reconstructions. Typically, a checking procedure like a heartbeat can check whenever the connection with a database is available and whether the corresponding module can execute its task. Moreover, no replica of data needs to be created because of subsequent attempts of writing. Then, all the data that are no longer locally useful have to be canceled to free the memory of the computer that hosts the NNs.

Finally, we discuss the updating of the NN characteristics. Typically, this is required on a longer time scale than the interval between two reconstruction cycles. Several reconstruction cycles may be required daily, while an update of the neural networks parameters could be needed only after some weeks, or even after some months. These different time scales and the importance of checking the generalization capability of the NNs may make one decide to avoid an automatic training procedure. In such a case the same set of NNs could be trained off line. Then, the folder containing the on line neural parameters could be replaced with the new ones.

Such a solution has actually been implemented for the Italian Sea Measurements Systems. The neural algorithms have been coded in C, while the other procedures have been developed in Perl. The PC used to train and run the neural networks is a PC, with a PENTIUM processor, connected to the network of the Measurements Systems and so to the central database.

8

Application of Approximation Theory and ARIMA Models

In this chapter we describe other algorithms as a possible alternative to artificial neural network (ANN) method for solving the reconstruction problem. Many algorithms, unlike ANN and simply NN, have been used for solving analogous problems. We selected two algorithms: the approximation operators which are a different version of ANN, already studied and explained in detail in Chapter 5, and the classical autoregressive integrated moving average (ARIMA) models widely used in the framework of time-series analysis. We apply both of them to our problem and we show with some examples that the ANN models have a much better performance.

8.0.1 Approximation Operator in One-Dimensional Case

In this section we will see the application of the approximation operators (5.15) (*Lenze operators*) to the significant wave height (SWH) data. We could not model the sea level (SL) time series with the Lenze's operator. In fact, as has been explained in Chapter 5, in order to apply the approximation operator for interpolating a gap consisting of h missing data, we must have a long time series with at least one point every h data. For the SL series this is impossible, as we often have gaps longer than one month. Thus, although the approximation operator is a valid interpolation method, it cannot estimate the SL series missing data.

In order to apply the Lenze operator to the SWH data, we have to first rewrite the generalized form of the operators generated by the sigmoidal functions, in the one-dimensional case:

$$\Omega^{(h)}(f)(x) = -\sum_{j\in\mathbb{Z}} \sigma\Big(j + \frac{1}{2} - \frac{x}{h}\Big)\Delta_f[hj, h(j+1)], \tag{8.1}$$

where from the definitions (5.8), (5.6) and (5.7):

$$\Delta_f[hj, h(j+1)] := \sum_{x\in C_j} (-1)^{\gamma(x,hj)} f(x) \tag{8.2}$$

$$C_j = \{hj, h(j+1)\} \tag{8.3}$$

$$\gamma(hj, hj) = 1, \quad \gamma(h(j+1), hj) = 0, \tag{8.4}$$

and from (8.3), (8.4) it follows:

$$\Delta_f[hj, h(j+1)] = f(h(j+1)) - f(hj). \tag{8.5}$$

Finally, the approximation operator can be rewritten in the one-dimensional case as follows:

$$\Omega^{(h)}(f)(x) = -\sum_{j\in\mathbb{Z}} \sigma\Big(j + \frac{1}{2} - \frac{x}{h}\Big)\Big(f(h(j+1)) - f(hj)\Big). \tag{8.6}$$

In order to reconstruct the SWH series of the Monopoli station, two main step functions have been applied. First a sigmoidal function has been chosen equal to

$$\sigma(\xi) := \frac{1}{2}(1 + \tanh(\xi)), \quad \xi \in R, \tag{8.7}$$

and then the same sigmoid function defined in the NN application:

$$\sigma(\xi) := \frac{1}{(1 + e^{-\lambda\xi})}, \quad \xi \in R. \tag{8.8}$$

In Section 8.0.2 we show the results obtained from the application of Lenze's operator to a time series of SWH.

8.0.2 Approximation for the SWH

The approximation operator (5.15) can be used as an interesting interpolation method to fill the gaps of sea time series. In this section we show the reconstruction obtained by applying this model to the significant SWH series of the Monopoli station, the same station to which we applied the NN algorithm.

First of all we have to decide the value of the *interpolation step* h. The algorithm is efficient only if a suitable *interpolation step* has been defined. This is an important choice. In fact, it is possible to use the approximation operator only if the actual data have a distribution without gaps in the interpolation points. See, for example, the case in which the operator $\Omega^{(h)}(f)(x)$

$$\Omega^{(h)}(f)(x) = -\sum_{j\in\mathbb{Z}} \sigma\Big(j + \frac{1}{2} - \frac{x}{h}\Big)\Big(f(h(j+1)) - f(hj)\Big) \tag{8.9}$$

has an interpolation step $h = 8$, corresponding to the time interval of one day. A step equal to 8 requires a long time series in which we have a data entry every 24 hours. The high frequency of gaps in the SWH series, in particular during sea storms, complicates the possibility of applying this operator. To overcome this problem we have applied Lenze's operator in the one-dimensional case to the subsequence with the minimum of missing data, the SWH time series of the Monopoli station. The algorithm has been employed with a different interpolation step; first with $h = 8$ (Fig. 8.1 and 8.2) and then with a step $h = 16$ (Fig. 8.3 and 8.4). In each case the number N of data is equal to 1600.

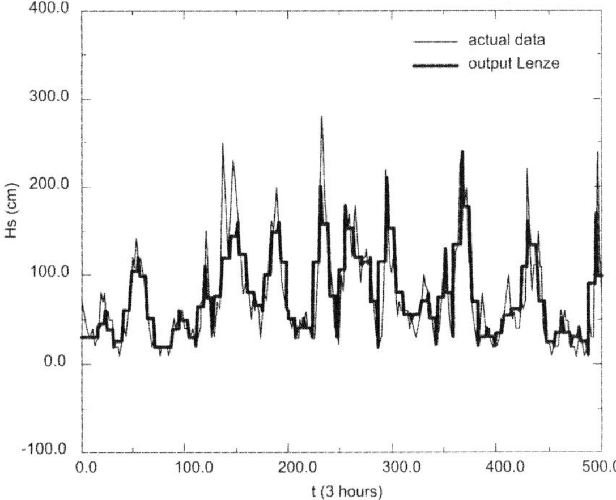

Fig. 8.1. First application to a subset of the significant height series of the Monopoli station with interpolation step $h = 8$. MAD is $LE = 21.7$ cm.

Analogously to the NN algorithm the error function has been defined as follows:

$$LE = \frac{1}{K} \sum_{i \neq jh}^{N} |O[i] - R[i]|, \quad j = 1, \dots, N/h, \tag{8.10}$$

where K indicates the number of actual data reconstructed by the algorithm, so $K = N - N/h$, $O[i]$ is the observed value and $R[i]$ is the value estimated with

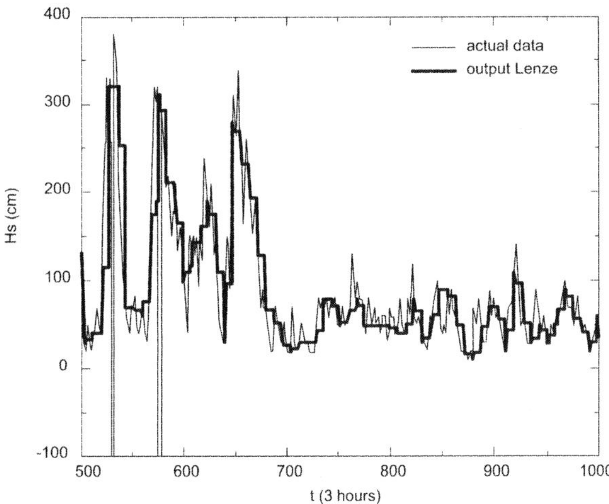

Fig. 8.2. Second application to a subset of the significant height series of the Monopoli station with an interpolation step $h = 8$. MAD is $LE = 21.7$ cm.

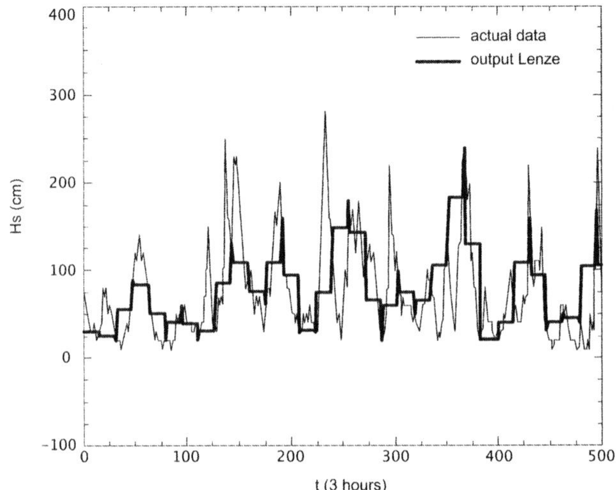

Fig. 8.3. Application to the same subset of Fig. 8.1 with a different step, $h = 16$. MAD is $LE = 27.8$ cm.

the approximation operator. In the application the error used in (8.10) is called MAD (mean absolute deviation). It is the same as the mean absolute error. This is a term used in probability for measuring the absolute deviation from the average of the values taken by a random variable, although here it is used with a different meaning. For $h = 8$, $N = 1600$ and choosing the sigmoidal function (8.8) the MAD $LE = 22.7$

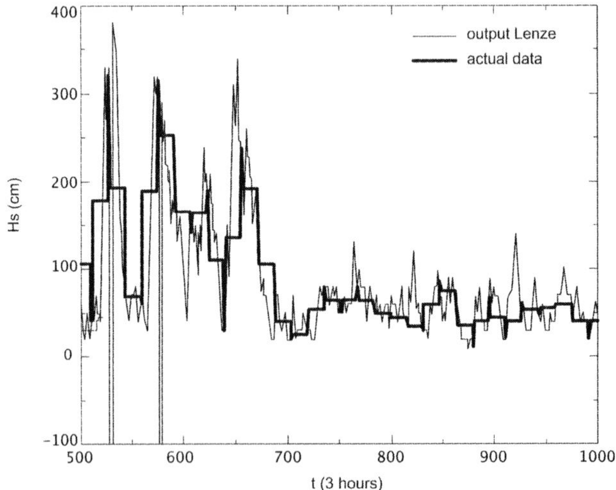

Fig. 8.4. Application to the same subset of Fig. 8.2 with a different step, $h = 16$. MAD is $LE = 27.8$ cm.

cm has been obtained. In Fig. 8.1 and 8.2 we can see the actual data of the SWH of the Monopoli station and the corresponding data modeled by the approximation operator with the parameters defined above. We used the approximation operator with different h. With $h = 16$, $N = 1600$ and the sigmoidal function (8.8), the MAD produced by the algorithm is $LE = 27.8$ cm (Fig. 8.3 and 8.4). From Fig. 8.1 and 8.2 it is possible to note that the Lenze operator approximates the Monopoli SWH time series with a function similar to a step function. For this reason we have decided to stop the tests at the interpolation step $h = 16$.

Although the error obtained with the interpolation step $h = 8$ is similar to the error obtained with the NN during the testing phase, the reconstructions are not comparable. In fact the output of the Lenze operator does not follow the trend of the actual data and does not reproduce the highest waves, thus omitting the most interesting part of the SWH series. Therefore, the results of the Lenze's operator become unreliable when we apply it to a sea storm period. It is known that SWH is for the construction of the harbors, and for this reason there is great interest in sea storm data. It is very important to evaluate the risk of sea storm impact on a structure, and so we cannot use a model that fails during a sea storm.

In conclusion, the results obtained applying the approximation operator to the SWH series are not as satisfactory as the results obtained using the NN.

8.1 ARIMA Models

The theoretical development and the simple formalism of ARIMA models makes this methodology useful for the application to the SWH and SL time series reconstruction. Anyway, this is not a purely predictive problem, since the gaps have a beginning and a proper end, and the reconstruction of the missing measurements should also take into account the data after the gap. ARIMA models, on the other hand, provide estimates using either the previous values or the following values. Furthermore, ARIMA prediction properties imply the inability of the models to predict extreme events adequately. In any case, the strong limit of the application of ARIMA in such a problem concerns the forecast error. If we define

$$\mathcal{E}_t(j) = x_{t+j} - E_t x_{t+j}$$

as the forecast error at the jth data, it can be proved that it is unbiased (i.e., $E_t\mathcal{E}_t(j) = 0$), but its variance is an increasing function of the number of iterations, leading to inaccurate predictions. As $j \to \infty$ the variance of the forecast error tends towards a constant value of the same order of magnitude as the sequence's unconditional variance $\{x_t\}$. Thus, unless the elementary error is really small, the reconstruction of long gaps does not function well, as we show in this section.

Following the Box–Jenkins methodology [10], it is possible to express the ARIMA model for a stationary time series x_t as

$$x_t = \varphi_1 x_{t-1} + \cdots + \varphi_p x_{t-p} + a_t - \theta_1 a_{t-1} - \cdots - \theta_q a_{t-q}, \tag{8.11}$$

where $\varphi_1, \dots, \varphi_p, \theta_1, \dots, \theta_q$ are parameters that must be estimated from the data, $a_t, \dots, a_{t-q}$ are i.i.d. random variables with zero mean and variance one (white noise, WN in the following), and $x_{t-1}, \dots, x_{t-p}$ are time series values at times $t-1, \dots, t-p$. The numbers p, q define the orders of the model and, in addition, there is an integer parameter d that establishes the number of differentiations of the original series in order to obtain a stationary x_t (see the following sections). By differentiation ∇ we mean the following operation on the data:

$$\nabla x_t = x_t - x_{t-1}.$$

Therefore, equation (8.11) represents an ARIMA(p, d, q) model.[1]

There are three fundamental phases in the estimation of an ARIMA model: *identification*, *estimation* and *prediction*. In the first the goal is to find the correct values of p, d, q; the second regards the building of the model, estimating all the coefficients from the data. Finally, the last phase produces a reconstruction of the gap, using the prediction properties of the ARIMA model. The routines of the statistical software package SAS and, in particular, the PROC ARIMA routine are used to optimize the above-mentioned three phases.

8.1.1 Preliminary Analysis on SHW Measures

Different sets of SWH measures from Crotone (CR), Mazara del Vallo (MA), Monopoli (MO) and Ponza (PO) stations (see Fig. 8.5–8.8) in the period January 1, 1989–December 31, 1998 have been considered for the application of ARIMA models. The negative values represent missing data. The basic statistics for each station are listed in Tables 8.1, 8.2 and 8.3. The symbols μ, σ and σ^2 represent the mean value, the standard deviation and the variance respectively. The first two indicators

[1] In the frame of a more compact formalism, the orders p, d, q represent the number of times we apply three fundamental operators to the original series. These operators are the *autoregressive operators*,

$$\varphi^p(B) = 1 - \varphi_1 B - \varphi_2 B^2 - \dots - \varphi_p B^p,$$

the *prime difference operator*,

$$\nabla^d = (1 - B)^d,$$

and the *moving average operator*,

$$\theta^q(B) = 1 - \theta_1 B - \theta_2 B^2 - \dots - \theta_q B^q,$$

where $Bx(t) = x(t-1)$. In this way the ARIMA(p, d, q) model is expressed by the equation

$$\varphi^p(B)\nabla^d x_t = \theta^q(B) a_t.$$

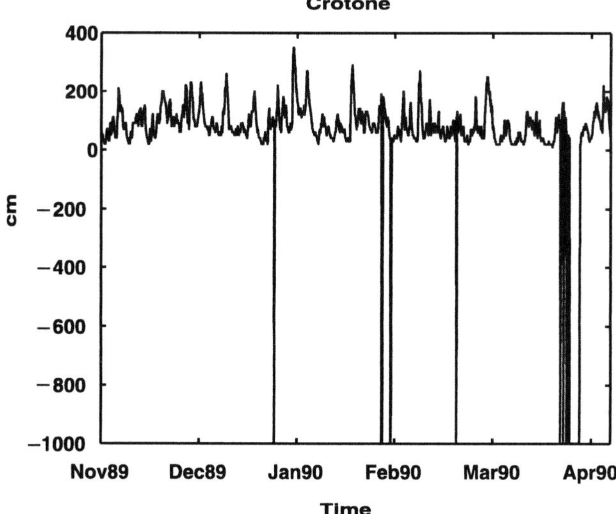

Fig. 8.5. SWH measures of Crotone.

are measured in centimeters (cm) while the last is measured in square centimeters (cm^2).

Among the considered stations, Monopoli seems to be more complete for quantity of information, number of gaps and length of the gaps interval. Therefore, it constitutes the benchmark for estimating ARIMA model and verifying its performance.

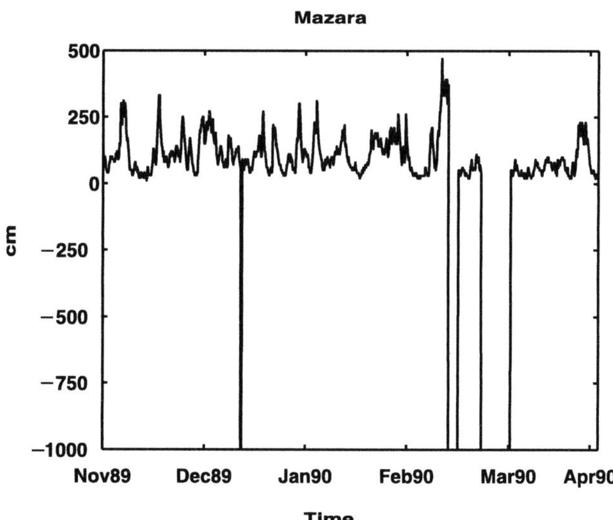

Fig. 8.6. SWH measures of Mazara del Vallo.

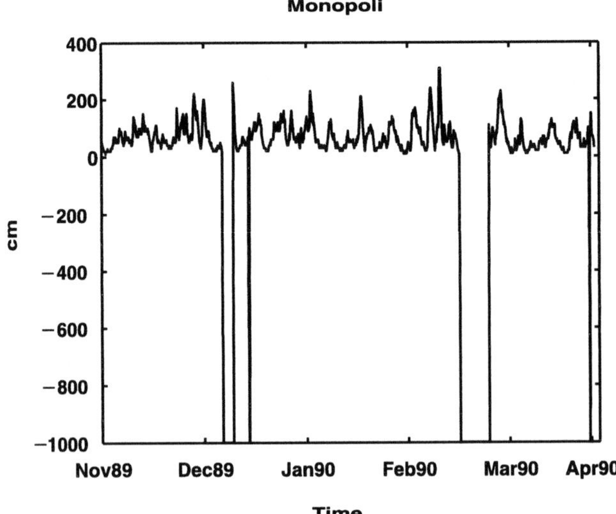

Fig. 8.7. SWH measures of Monopoli.

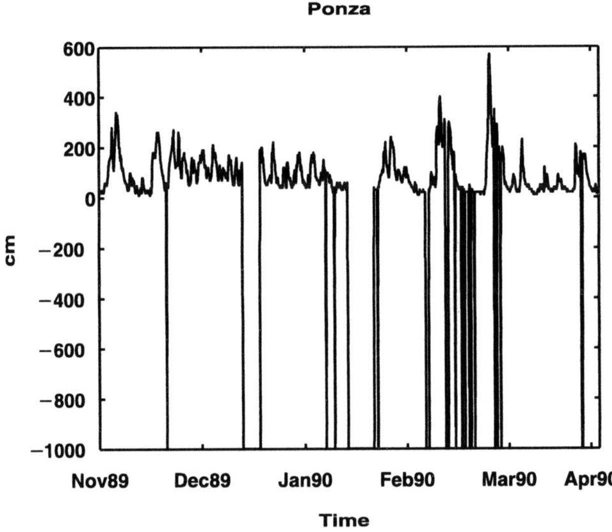

Fig. 8.8. SWH measures of Ponza.

8.1.2 Sampling

Considering that a single missing measure presents no difficulty, it is possible to focus on the first two extended gaps in the Monopoli SHW, occurring at August 15–16, 1989 (six missing data) and at December 7–10, 1989 (twenty six missing data), respectively. The size of these gaps is sufficient to evaluate the performance of an

Table 8.1. Basic statistics on SWH measures of the selected stations.

Station	N	μ	σ	σ^2	Max	Min	Range
CR	27,245	72.91	62.64	3923.34	560	7	553
MA	24,900	98.25	72.16	5206.77	590	0	590
MO	27,411	68.05	52.78	2785.98	510	8	502
PO	26,444	85.56	69.57	4840.06	580	7	573

Table 8.2. Basic statistics on gaps (number of gaps per station, number of missing measures per station, statistics on gap's length).

Station	N	Missing	μ	σ	Mode	Median	Max	Min
CR	535	1971	3.7	18.5	1	1	360	1
MA	968	4261	4.4	13.8	1	1	249	1
MO	373	1805	4.8	13	1	1	179	1
PO	463	2772	5.99	16.5	1	2	264	1

Table 8.3. Basic statistics on length of continuous subsequences in the selected stations.

Station	μ	σ	Mode	Median	Max	Min
CR	50.83	121.7	7	17	1270	1
MA	25.7	51.15	1	7	510	1
MO	73.3	155.8	1	18	1367	1
PO	57	124.15	1	14	1632	1

ARIMA model and, to this end, the chosen samples for estimation are the sequences of measures before the selected gap. The first sequence, related to the first gap and called MONOP1, begins at July 1, 1989; the second, called MONOP2, begins at September 5, 1989. (See Fig. 8.9 and 8.10.)

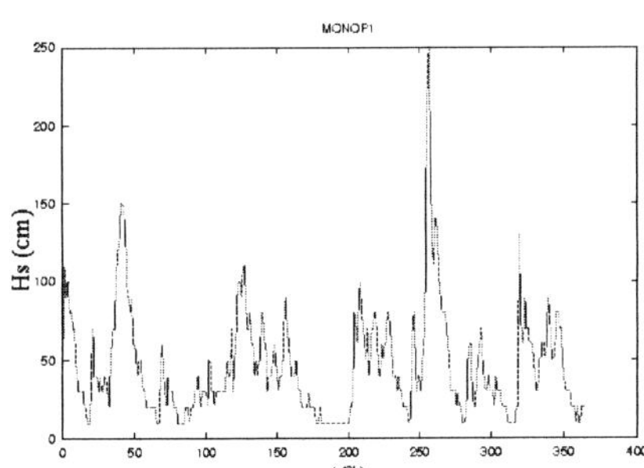

Fig. 8.9. MONOP1 sequence: 12 p.m. July 1, 1989–12 a.m. August 15, 1989

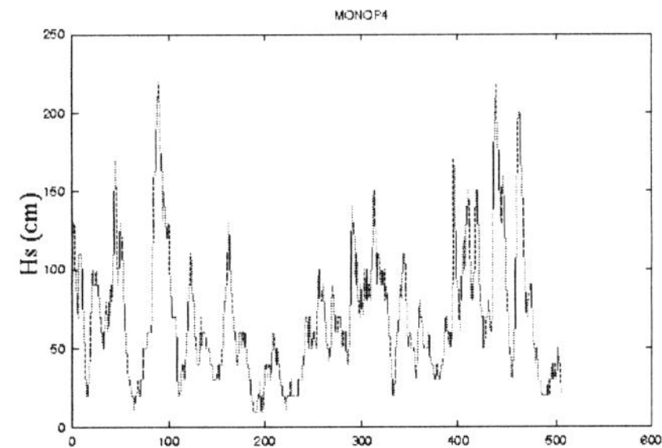

Fig. 8.10. MONOP2 sequence: 9 a.m. October 5, 1989–3 p.m. December 7, 1989

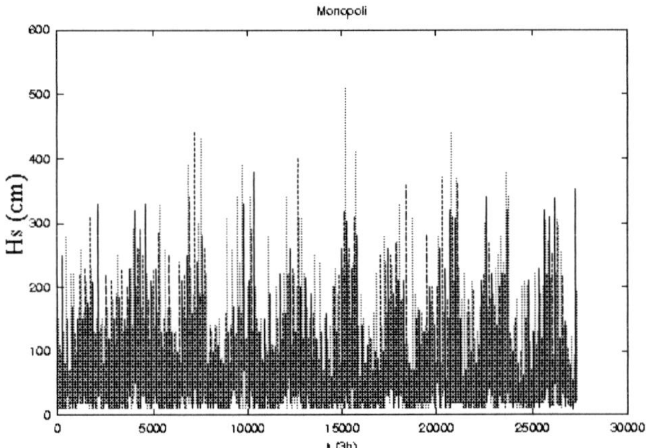

Fig. 8.11. MONOP3: Monopoli series 1989–1999 without gaps.

According to the requirement that the estimation of the model should also take into account information regarding the measures just after the gaps, it is important to consider as a further sample the entire sequence of measures deprived of the gaps. Therefore, we transformed the original interrupted series to a series without gaps (see Fig. 8.11).

The three selected sequences are standardized with the mean and standard deviation of the entire sequence (see Table 8.4).

Table 8.4. Basic statistics of the samples extracted from Monopoli.

Sequence	N	μ	σ	σ^2	Min	Max	Range
MONOP1	365	46.52	34.42	1184.84	10	250	240
MONOP2	507	69.47	41.13	1692.01	10	220	210
MONOP3	27,411	68.05	52.78	2785.98	510	8	502

8.1.3 Identification of the Model

The first objective is to verify the stationarity of the time series. This can be done by studying the autocorrelation for the three sequences or, a posteriori, checking the behavior of residues (i.e., the difference between predicted and observed values) that must be of the WN type.[2]

The autocorrelation plots (Fig. 8.12–8.17) show a more rapid decay of correlations in the case of the differentiated series. For the original series, the correlation

[2] This follows from the hypothesis that the ARIMA estimate of the process admits a statistical error distributed as a Gaussian with zero mean and standard deviation corresponding to the average experimental error.

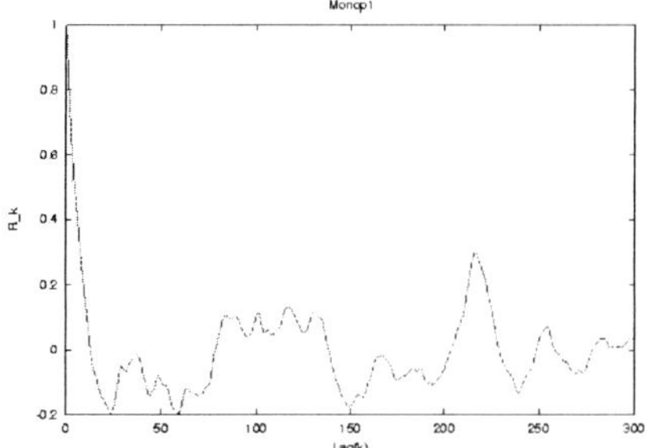

Fig. 8.12. Autocorrelation of the standardized series MONOP1.

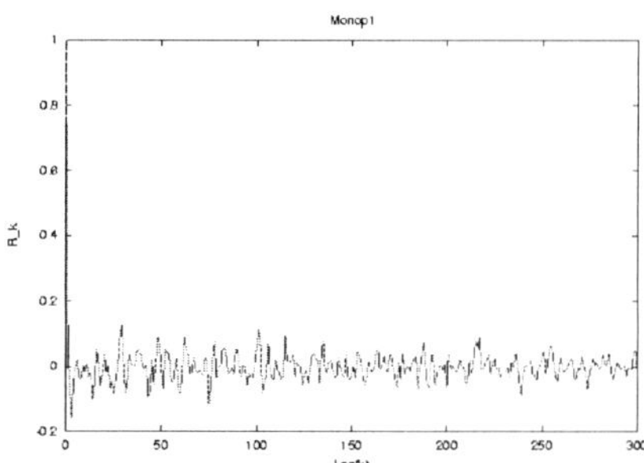

Fig. 8.13. Autocorrelation of the first difference of MONOP1.

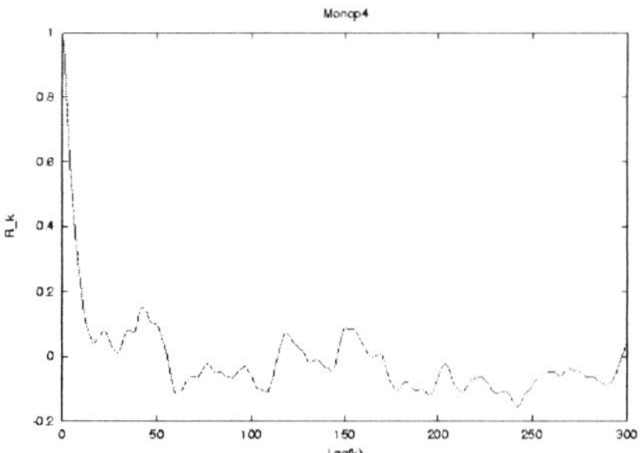

Fig. 8.14. Autocorrelation of the standardized series MONOP2.

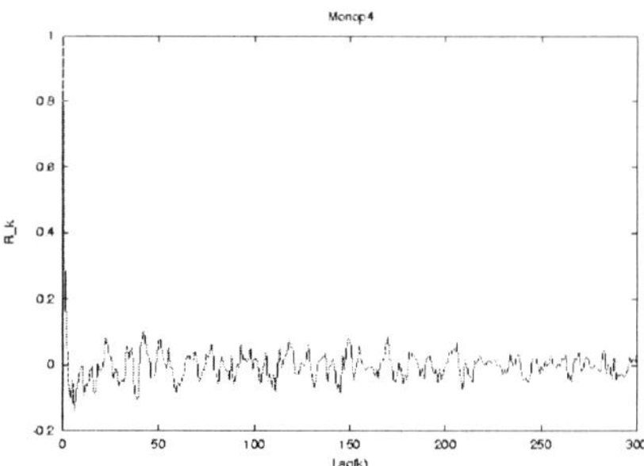

Fig. 8.15. Autocorrelation of the first difference of MONOP2.

has an oscillating behavior that clearly expresses non-stationarity. In contrast, the first difference series has an autocorrelation that has a rapid decrease and, after a few lags, it fluctuates around zero within the WN band of confidence.[3] The stationarity of the differentiated series gives us the important information that the value of the differences order is $d = 1$. The methods ESACF (extended sample autocorrelation function) and SCAN (Smallest CANonical correlation), proposed by Tsay and Tiao [72], have been used in order to determine the autoregressive order(p) and the moving average order (q). The first method considers the matrix of autocorrelation of trial models constructed for certain values of the orders p, q, while the second deals with the eigenvalues (the canonical correlations) of the correlation matrix among trial models. Both methods require optimization algorithms and can be applied to stationary and non-stationary time series, estimating $p + d$ and q values. The preliminary estimates for the three series emphasize different possible values for p and q (Tables 8.5–8.7). Their correct selection refers to a simplicity criterion concerning the number of parameters of the model, together with the verification of the convergence

[3] This is a standard procedure. The band of confidence is usually defined as $\pm\sigma/\sqrt{N}$ around the zero value and it represents the band within which the signal can be considered negligible.

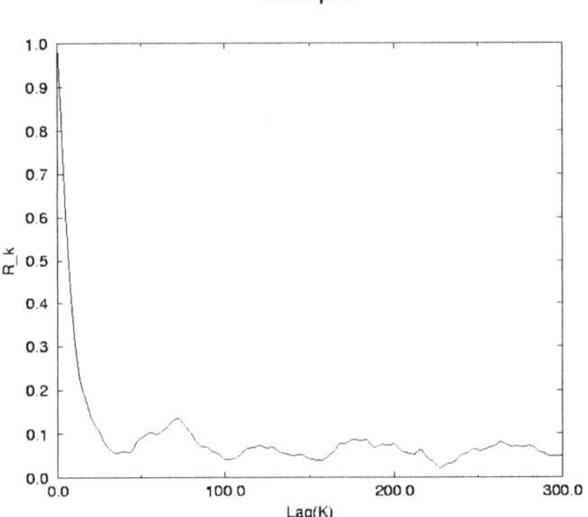

Fig. 8.16. Autocorrelation of the standardized series MONOP3.

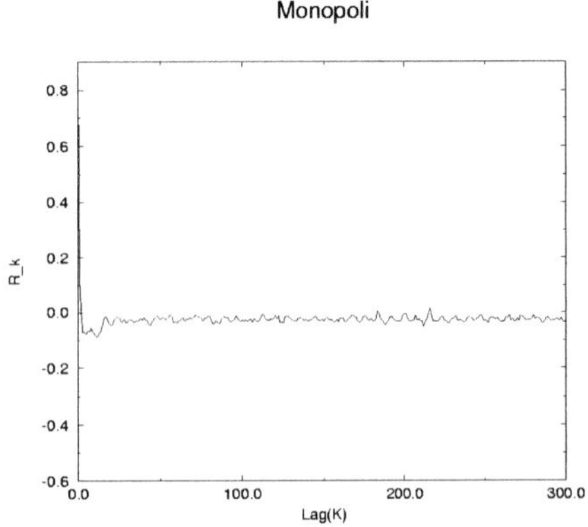

Fig. 8.17. Autocorrelation of the first difference of MONOP3.

Table 8.5. Orders p, q for MONOP1.

SCAN		ESACF	
$p+d$	q	$p+d$	q
2	1	2	2
1	2	0	3
3	0	1	3
0	3	6	5
		8	5

Table 8.6. Orders p, q for MONOP2.

SCAN		ESACF	
$p+d$	q	$p+d$	q
4	2	1	6
1	4	2	6
6	1	3	6
		4	6
		6	7
		7	7
		8	7
		9	6

Table 8.7. Orders p, q for MONOP3.

SCAN		ESACF	
$p+d$	q	$p+d$	q
3	3	6	13
2	4	9	15
0	29	10	15
30	0	11	15
		12	15
		13	15
		28	27
		30	25
		0	29
		1	29
		2	29
		3	29

Table 8.8. Identified models.

Series	Models
MONOP1	ARIMA(0, 1, 3)
MONOP2	ARIMA(1, 1, 4)
MONOP3	ARIMA(0, 1, 29)

Table 8.9. Estimated parameters for MONOP1.

Parameter	Estimated value	Std Error	t	Approx Pr > $\|t\|$	Lag
μ	-1.9326×10^{-3}	0.01407	−0.14	0.8908	0
θ_1	−0.08609	0.05181	−1.66	0.0975	1
θ_2	0.06810	0.05178	1.32	0.1893	2
θ_3	0.15212	0.05188	2.93	0.0036	3

of the algorithms.[4] It can be noted how the three tables do not show any common pair p, q, implying the choice of different models for different gaps, in the case of MONOP1 and MONOP2.

8.1.4 Estimate of the Model

On the basis of the results of the previous section, three models have been detected, one for each considered sample (Table 8.8, Fig. 8.18 and 8.19). Following the results of the previous exploration, the estimation of the models requires the application of an optimization procedure (generally a least square algorithm) in order to determine the parameters φ, θ of formula (8.11) and to perform the analysis of residues, for which a WN behavior is expected.

The results are summarized in Tables 8.9–8.13. In Tables 8.9 through 8.11 we show the estimated values of the parameters with the standard error associated to each of them, the t statistics with the associated probability and the lag corresponding to each parameter. The t value, in particular, refers to the significance of the corresponding term in the ARIMA model, with respect to the average of the time series. If the component modeled by a term is far from the average then it is significant and the t value associated to the parameter is far from a critical value. This distance is expressed by the level of probability in the next column: a lower value of probability corresponds to a higher value of t and greater significance of the parameter.

Observing Tables 8.12 and 8.13 it can be noted that the correlation among parameters never exceeds 0.6 in MONOP2 and 0.1 in MONOP1.[5] As far as the entire

[4] Sometimes both methods predict p and q values, although the optimization algorithm could have not reached a minimum of the cost function.

[5] By correlation between two parameters we mean the Pearson cross-correlation between the two sets of values assumed by the parameters during their estimation procedure. In this case it expresses a measure of the independence of the parameters and therefore of

Table 8.10. Estimated parameters for MONOP2.

Parameter	Estimated value	Std Error	t	Approx Pr > $\|t\|$	Lag
μ	-1.5297×10^{-3}	2.2568×10^{-3}	−0.68	0.4982	0
θ_1	0.66140	0.05617	11.77	<0.0001	1
θ_2	0.09364	0.05276	1.78	0.0765	2
θ_3	0.17932	0.05278	3.40	0.0007	3
θ_4	0.04298	0.05050	0.85	0.3951	4
ϕ_1	0.86678	0.03535	24.52	<0.0001	1

Table 8.11. Estimated parameters for MONOP3.

Parameter	Estimated value	Std Error	t	Approx Pr > $\|t\|$	Lag
μ	5.27×10^{-6}	1.001×10^{-4}	0.05	0.9580	0
θ_1	−0.07333	0.00604	−12.14	<0.0001	1
θ_2	0.03825	0.00605	6.32	<0.0001	2
θ_3	0.09305	0.00606	15.36	<0.0001	3
θ_4	0.07915	0.00608	13.01	<0.0001	4
θ_5	0.09695	0.00610	15.89	<0.0001	5
θ_6	0.07355	0.00613	11.99	<0.0001	6
θ_7	0.07966	0.00615	12.96	<0.0001	7
θ_8	0.05805	0.00617	9.41	<0.0001	8
θ_9	0.06693	0.00617	10.84	<0.0001	9
θ_{10}	0.06553	0.00619	10.59	<0.0001	10
θ_{11}	0.07269	0.00620	11.73	<0.0001	11
θ_{12}	0.06447	0.00621	10.38	<0.0001	12
θ_{13}	0.05564	0.00622	8.94	<0.0001	13
θ_{14}	0.03913	0.00623	6.28	<0.0001	14
θ_{15}	0.02491	0.00624	3.99	<0.0001	15
θ_{16}	0.00114	0.00623	0.18	0.8553	16
θ_{17}	−0.00408	0.00622	−0.66	0.5118	17
θ_{18}	0.01247	0.00621	2.01	0.0448	18
θ_{19}	0.01635	0.00620	2.64	0.0083	19
θ_{20}	0.02890	0.00619	4.67	<0.0001	20
θ_{21}	0.02265	0.00617	3.67	0.0002	21
θ_{22}	0.00948	0.00617	1.54	0.1244	22
θ_{23}	−0.00445	0.00615	−0.72	0.4693	23
θ_{24}	−0.00433	0.00613	−0.71	0.4803	24
θ_{25}	0.00076	0.00610	0.12	0.9011	25
θ_{26}	0.01203	0.00609	1.98	0.0481	26
θ_{27}	0.00557	0.00606	0.92	0.3578	27
θ_{28}	0.00179	0.00606	0.30	0.7679	28
θ_{29}	0.02289	0.00604	3.79	0.0002	29

the estimated terms of the model. A correlation as low as possible among the estimated parameters is expected.

Table 8.12. Correlation among parameters (MONOP1).

Corr	μ	θ_1	θ_2	θ_3
μ	1.000	−0.003	−0.001	−0.002
θ_1	−0.003	1.000	0.088	−0.056
θ_2	−0.001	0.088	1.000	0.077
θ_3	−0.002	−0.056	0.077	1.000

Table 8.13. Correlation among parameters (MONOP2).

Corr	μ	θ_1	θ_2	θ_3	θ_4	ϕ_1
μ	1.000	−0.028	−0.003	0.001	0.014	−0.048
θ_1	−0.028	1.000	−0.420	−0.105	−0.435	0.612
θ_2	−0.003	−0.420	1.000	−0.446	−0.095	0.044
θ_3	0.001	−0.105	−0.446	1.000	−0.446	−0.064
θ_4	0.014	−0.435	−0.095	−0.446	1.000	−0.475
ϕ_1	−0.048	0.612	0.044	−0.064	−0.475	1.000

sequence is concerned, MONOP3, the matrix is not reported, but the correlation evaluated is always lower than 0.02.

At this point it is important to attempt a comparison between the model estimated from the entire data sequence (MONOP3) and the two others (MONOP1 and MONOP2), coming from interrupted series. Actually, even if all these models are evaluated by the same methodology, the approach is substantially different. The MONOP1 model, like the MONOP2 one, assumes that all the information needed to reconstruct the gap is contained in the measures that precede it. However, the model for MONOP3 also considers the information contained in the values following the gap.

The models obtained for MONOP1 and MONOP2 are clearly more stable than the one estimated for MONOP3, considering the great number of parameters it involves. It seems, however, that such parameters have the minimum correlation and so they span the information more effectively. The t values in Table 8.11 show this property.

8.1.5 Prediction: Finding Missing SHW Values

As usual, the reconstruction of the missing measures requires a preliminary check of the adaptation of the models estimated from real data. For an ARIMA model, this implies the comparison of the process

$$x_t = \hat{\varphi}_1 x_{t-1} + \cdots + \hat{\varphi}_p x_{t-p} + a(t) - \hat{\theta}_1 a_{t-1} - \cdots - \hat{\theta}_q a_{t-q}$$

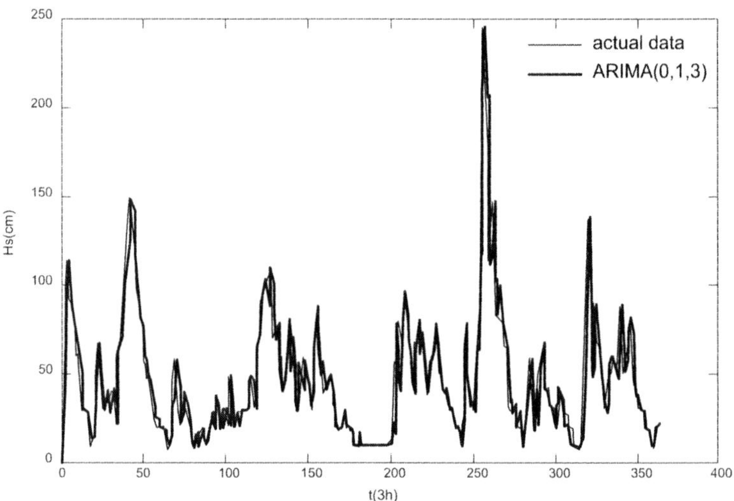

Fig. 8.18. Comparison of ARIMA(0,1,3) with MONOP1.

with the values of $X(t)$ used for estimating the model ($\hat{\varphi}, \hat{\theta}$ are the estimated parameters). To this end, we will consider the error[6]

$$\mathcal{E} = \frac{1}{P} \sum_{t=1}^{P} (x_t - X(t))^2 , \tag{8.12}$$

which is the average sum of squares of 1-step residues. For the sequences MONOP1 and MONOP2 the square root of the error is $\sqrt{\mathcal{E}} = 10$ cm, while for the complete sequence MONOP3, $\sqrt{\mathcal{E}} = 12$ cm. The error levels are clearly very low, but this does not say anything about prediction of the missing measures. What we can observe from this comparison is the correct convergence of the estimating algorithm. Regarding the estimate of the missing values of the gaps, the reference indicator is the significance of the prediction itself, which is expressed by the confidence interval at a probability level of 0.95. This, by definition, is increasing non-linearly with the steps of the prediction and, even for small steps, can be very wide as shown by Fig. 8.20–8.25.

We should note a substantial homogeneity among the predictions obtained by the "interrupted series models" (MONOP1, MONOP2) and those obtained by the model constructed on the complete sequence (MONOP3), when applied to the same gaps. Whatever the case, the real problem of ARIMA models seems to be their increasing prediction error.[7] At this level it is extremely important to verify the ability of this

[6] This kind of error is not usual for the ARIMA model, but it is introduced in order to make a comparison with the other methodology. In fact, it corresponds to the learning error LE introduced in Chapter 4.

[7] As an elementary example of this behavior, let us consider a simple AR(1) model $x_t = \phi_0 + \phi_1 x_{t-1} + a_t$. It follows that

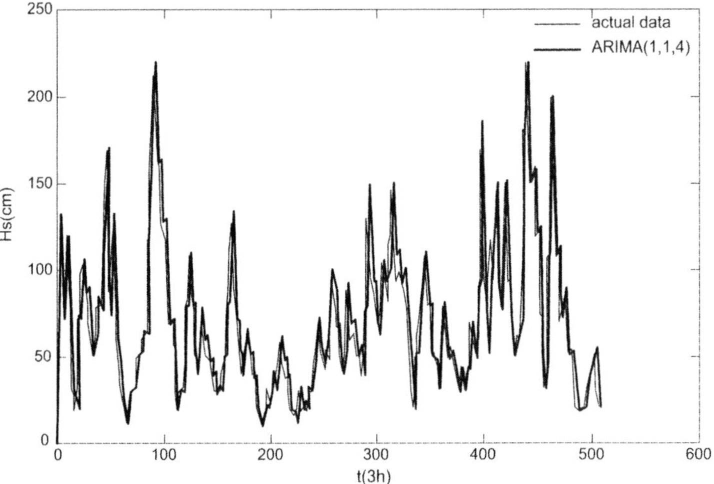

Fig. 8.19. Comparison of ARIMA(1,1,4) with MONOP2.

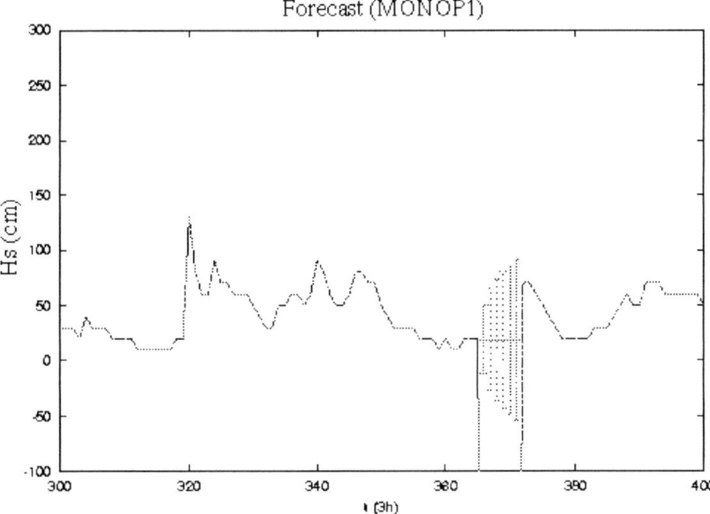

Fig. 8.20. Filling of the gap following MONOP1, from 3 p.m. August 15, 1989 to 6 a.m. August 16, 1989 (6 missing measures).

$$Ex_{t+j} = \phi_0 \left(1 + \phi_1 + \phi_1^2 + \cdots + \phi_1^{j-1}\right) + \phi_1^j x_t;$$

therefore, the forecast error is

$$\begin{aligned}\mathcal{E}_t(j) &= x_{t+j} - E_t x_{t+j} \\ &= a_{t+j} + \phi_1 a_{t+j-1} + \phi_1^2 a_{t+j-2} + \phi_1^3 a_{t+j-3} + \cdots + \phi_1^{j-1} a_{t+1},\end{aligned}$$

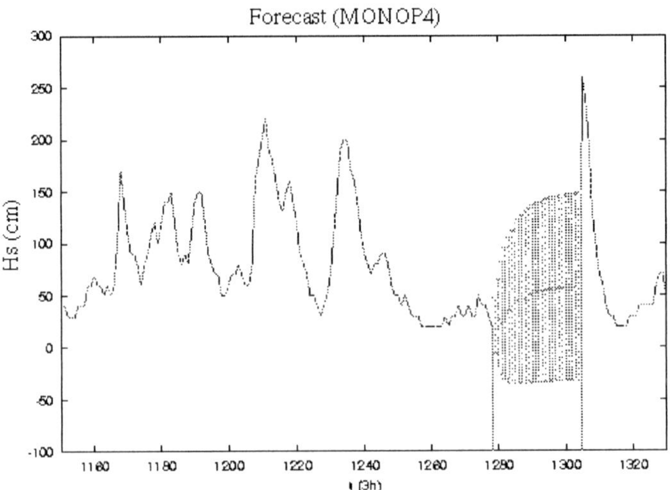

Fig. 8.21. Filling of the gap following MONOP2, from 6 p.m. December 7, 1989 to 9 p.m. December 10, 1989 (26 missing measures).

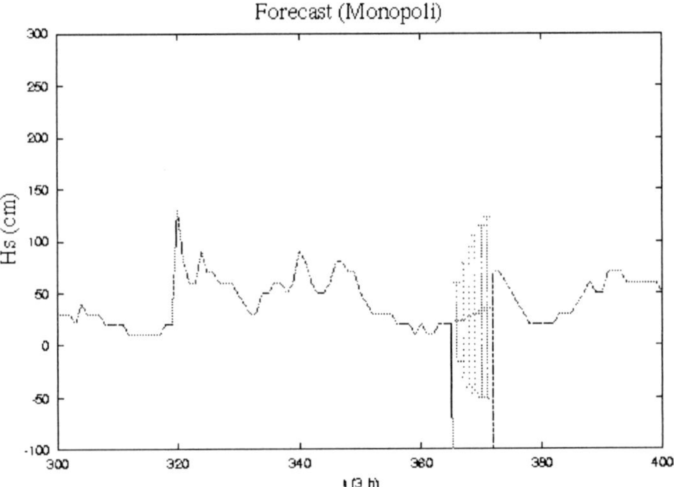

Fig. 8.22. Filling of the gap following MONOP1 by means of ARIMA(0,1,29), estimated on MONOP3.

---

which is unbiased. If we assume that the stochastic components of the model $\{a_t\}$ are independent with variance σ^2, the variance of the error is

$$\text{var}\,(\mathcal{E}_t(j)) = \sigma^2 \left[1 + \phi_1^2 + \phi_1^4 + \phi_1^6 + \cdots + \phi_1^{2(j-1)}\right]$$

expressing the increasing dependence on j. In the limit as $j \to \infty$, the variance converges to $\sigma^2/(1-\phi_1^2)$, that is, the unconditional variance of the $\{x_t\}$ sequence.

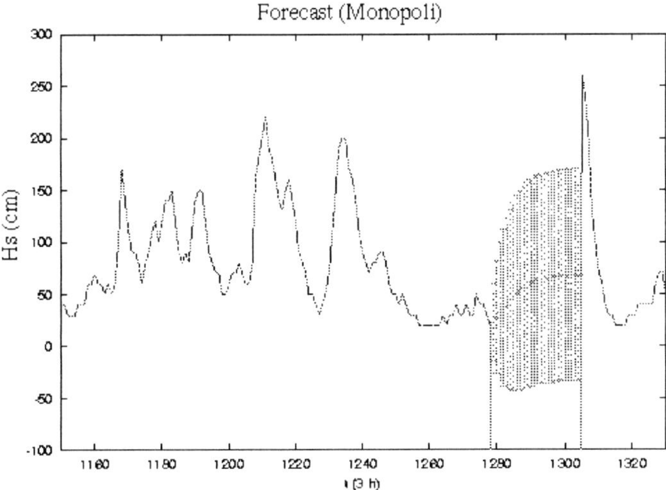

Fig. 8.23. Filling of the gap following MONOP2 by means of ARIMA(0,1,29), estimated on MONOP3.

methodology to approach the problem of missing measures with respect to the NN. The comparison seems to be complex from the methodological point of view, since ARIMA models produce a homogeneous response to the input information (the data are of the same station), while the NN application considered in this book uses input data of different but correlated stations. It is possible to compare the 1-step error of

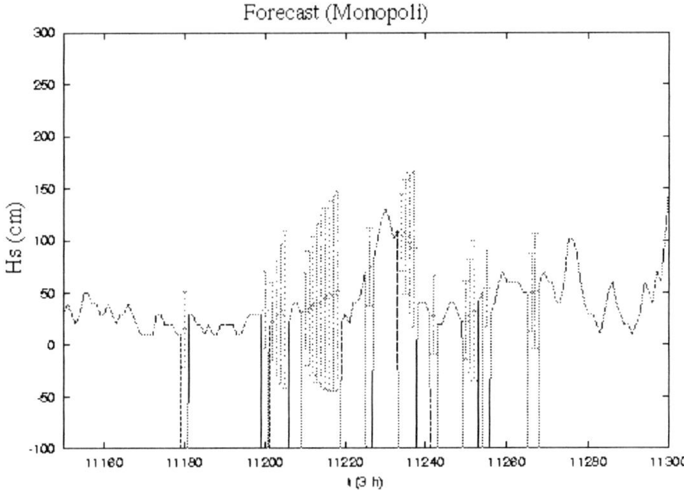

Fig. 8.24. Filling of the gaps from 6 p.m. April 24, 1993 to 9 p.m. May 12, 1993 of MONOP3, by means of ARIMA(0,1,29)(I).

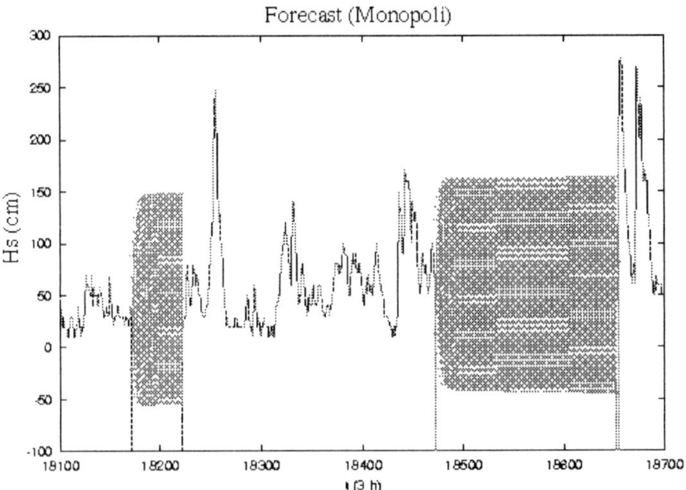

Fig. 8.25. Filling of the gaps from 12 a.m. September 10, 1995 to 6 p.m. November 24, 1995 of MONOP3, by means of ARIMA(0,1,29)(II).

equation (8.12) with the *LE* of NNs and to make a comparison of the confidence interval with the *GE*.

Following this criterion, it can be noted that, even if the *LE* is smaller for ARIMA than for NNs, for the prediction of missing measures, the confidence interval associated to ARIMA models has an order of magnitude of a typical value of the data. This happens even in the case of the prediction of a single measurement (see Fig. 8.20). For gaps of medium or long duration, the amplification of the error, typical of such models, necessitates the extension of the confidence interval in order to maintain the level of significance at 95%. Unfortunately, this means that at the last missing measure the interval is so wide as to be comparable with an extreme fluctuation of the data (see Fig. 8.21–8.25). It must be emphasized, moreover, that sometimes the prediction of the last missing measure does not match the first successive known measure, which could be outside the confidence interval (as can be seen in Fig. 8.21 and 8.23).

In conclusion, let us make some observations on the application of ARIMA models to SHW measures:

- The estimate of the models on subsequences, extracted from the data just before the gaps as in MONOP1 and MONOP2, turns out to be problematic, since to each subsequence there corresponds a different model which cannot be exported to predict other gaps. Actually, the procedure of the estimate and the diversity of the subsequences guarantee the reliability of the prediction only for the successive measure.
- In terms of prediction and rebuilding of missing measures, ARIMA models permit reliable results only in the case of short gaps. As seen in the case for the models estimated to rebuild the gap from hour 3 p.m. August 15, 1989 to hour 6 a.m.

September 16, 1989 (six missing measure, models on MONOP1 and MONOP3), the prediction converges after a few steps to the mean value, making the result less significant (Fig. 8.20). Then, even for gaps with length less than 10 measures, the ARIMA response also turns out to be inadequate for following small data fluctuations. These models should be applied to gaps containing a single missing measure, but a simple interpolation could also be highly effective in this case.

- Regarding the proposed sampling, from the point of view of a possible automatization of the process, it would be preferable not to divide the series into subsequences, for the reasons expressed above. Instead, one should consider the total sequence of measures, as in the case of MONOP3, because this guarantees the uniqueness of the model and more information for the reconstruction of the missing measures.

9

Extreme-Event Analysis

In this chapter we discuss how we selected a model for describing the probability distribution of the extreme events of sea waves and how the model has been made to suit the structure of the data. We also discuss the model's consequences, such as the evaluation of an m-year return level. Also, we check whether the time series obtained by means of artificial neural network (ANN) reconstruction has the same extreme events probability distribution as the original one. We further discuss the definition of a special ANN method dedicated to the reconstruction of single selected extreme or large events.

9.1 Preliminary Analysis

In order to understand whether a statistical model is suitable for a specific case, preliminary analysis of the available data sets is required. After a model fitting has been done, a diagnosis of the results has to be performed to evaluate the efficiency of the corresponding outcomes. We already know that the significant wave height (SWH) time series has a seasonal behavior. However, this time series can be considered almost stationary within each season, implying that the main statistical properties of interest are approximately constant. It is also necessary to distinguish events coming from different directions. For example, in the seas around Italy, depending on the site, sea waves generated by a wind which usually blows from the north could be more intense than the ones generated by a wind which blows from the east. In this case, it is correct as well as useful to analyze such events separately.

Identifying possible trends over several years is a climatological issue. A good answer could be given by analyzing at least 30 consecutive years. We do not look into this issue, because our source of data does not cover such a long time period. We shall deal with data taken every three hours for periods of about fourteen years, as reported in Table 9.1.

Our aim, first of all, is to identify any seasonal sea storm behavior. Table 9.2 shows, for each site, the level of high alert as defined by APAT. The meaning of this

Table 9.1. Sea wave data.

Site	Start	End
Alghero	01/07/1989 00:00	30/06/2003 21:00
Catania	01/07/1989 00:00	30/06/2003 21:00
Crotone	01/07/1989 00:00	30/06/2003 21:00
La Spezia	01/07/1989 00:00	30/06/2003 21:00
Mazara	01/07/1989 00:00	30/06/2003 21:00
Monopoli	01/07/1989 00:00	30/06/2003 21:00
Pescara	01/07/1989 00:00	30/06/2003 21:00
Ponza	01/07/1989 00:00	30/06/2003 21:00

Table 9.2. Level of first attention.

Site	Threshold (m)
Alghero	3.0
Catania	2.0
Crotone	2.5
La Spezia	3.0
Mazara	2.0
Monopoli	2.0
Pescara	3.0
Ponza	2.5

level is that a sea wave storm is important only if the corresponding significant wave heights exceed the level of first attention.

It is useful to plot for each site all the available data over the given attention level on the same time frame of one year. For convenience, we have considered a window of one year extending from July to the following June; Fig. 9.1 shows the case of the Alghero site where monthly averages are plotted. It is possible to identify two main periods of values: we have called them Winter and Summer. The period of each season changes according to the site, but the length of the Winter turns out to be nearly half of the Summer one; further details can be found in Table 9.3.

Thereafter, principal sectors of wave directions have been identified by evaluating the empirical joint density distribution function of H_s and D_m. For each site, an accurate representation of this function has been obtained by dividing the range of H_s and D_m in cells of 50 centimeters and 15 degrees for H_s and D_m, respectively. Thus, the principal sectors are the intervals of the wave directions which are more populated; in Table 9.4 the resulting principal sectors are reported.

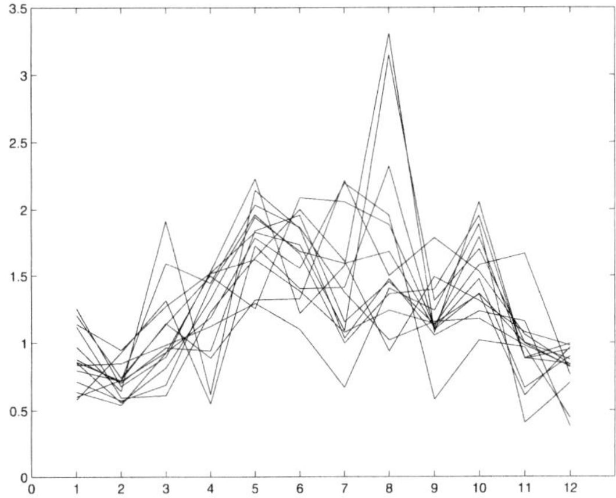

Fig. 9.1. Alghero's H_s (m), from July 1, 1989 to July 31, 2003: monthly averages for each year of measurements have been plotted. A year runs from July to the following June.

Table 9.3. Seasonal sets: a year runs from July to the next June.

Site	Winter	Summer
Alghero	Nov–Feb	Mar–Oct
Catania	Nov 15th–Mar	Apr-Nov 15th
Crotone	Nov–Apr 15th	Apr 15th–Oct
La Spezia	Oct 15th–Feb	Mar–Oct 15th
Mazara	Nov–Feb	Mar–Oct
Monopoli	Dec 15th–Apr	May–Dec 15th
Pescara	Nov–Apr 15th	Apr 15th–Oct
Ponza	Nov 15th–Feb	Mar–Nov 15th

Table 9.4. Principal sectors of wave directions

Site	I sector	II sector
Alghero	277.5 deg–322.5 deg	—
Catania	67.5 deg–122.5 deg	—
Crotone	112.5 deg–157.5 deg	352.5 deg–37.5 deg
La Spezia	187.5 deg–247.5 deg	—
Mazara	112.5 deg–157.5 deg	262.5 deg–307.5 deg
Monopoli	23.5 deg–67.5 deg	337.5 deg–22.5 deg
Pescara	337.5 deg–37.5 deg	—
Ponza	202.5 deg–246.5 deg	246.5 deg–292.5 deg

9.2 Model Fitting

Before attempting a fit by using a rather complex model, one should try using a simple one. Results could already be satisfactory, or further improvements could be made. In our case the events of interest belong to sea wave storms; thus, the estimation of the model will deal with the peak values of clusters of exceedances over a high threshold u. For such events a homogeneous Poisson-generalized Pareto distribution (GPD) model is suitable. Let us consider the main contents of this model:

- The number of observations occurring up to the time t constitutes the counting process

$$N(t) = \sum_{i=1}^{\infty} I(T_i \leq t), \quad t \geq 0$$

  that is a Poisson r.v. with parameter λ. Indeed, the probability of obtaining k observations up to the time t is given by

$$P\{N(t) = k\} = \frac{(\lambda t)^k}{k!} e^{-\lambda t}.$$

- Conditionally on $N \geq 1$, the observations over the threshold are outcomes from independent r.vs. identically distributed from a GPD, which has the following functional form:

$$G_{\xi,\sigma}(y) = \left[1 - \left(1 + \xi \frac{y}{\sigma}\right)^{-1/\xi}\right], \quad y \geq 0, \ \sigma > 0.$$

Bearing in mind the suggestions provided by the preliminary analysis, we will perform four different fits. In detail we will fit:

- the overall data sets without separating out any season or range of wave directions;
- the data within each principal range of wave directions, but without a seasonal characterization;
- the data within each season, but without a characterization of any principal range of wave directions;
- the data belonging to each season and each principal range of wave directions.

Obviously, the last two fits are equivalent for sites with only one principal sector of wave directions.

The model parameters are estimated maximizing the likelihood functions of the Poisson-GPD process on the available data. Performing several attempts, a sequence of sets of parameters can be identified. In order to choose the best one, the outcomes have to be carefully evaluated using proper diagnostic procedures. Graphical tools such as the Q-Q plot, P-P plot and residue scatter plot turn out to be very effective to this end; therefore, we will provide a short introduction to them.

First of all, we note that a quantile is defined as follows.

Definition 9.1 (Quantile) *Given a probability distribution function $F(x)$, the quantile is the generalized inverse function*

$$F^{\leftarrow}(t) = \inf\{x \in \mathbb{R} | F(x) \geq t\}, \quad \text{with } 0 < t < 1.$$

A Q-Q plot is easily built. Quantiles can be evaluated both on the given model for the probability distribution function as well as directly on the data set. Then, the former are plotted against the latter estimates. If the plotted points belong to the bisecting line within the confidence intervals, the proposed probability model turns out to be satisfactory.

By means of a P-P plot, probability values are compared. Again, the model agrees closely with the data if the points representing the couples of the empirical probabilities (evaluated directly from the available data) and the probabilities from the given fitted model are distributed along the bisecting line within the confidence intervals.

The diagnosis of the fitting models can be extended. Let us suppose that the first event of interest is observed at time T_0 and the following ones at times $T_1, T_2, \ldots, T_n$. Once the intensity of the Poisson process has been estimated, $\hat{\lambda}$, the residues can be computed as follows:

$$Z_k = (T_k - T_{k-1})\hat{\lambda}. \tag{9.1}$$

If data have actually been generated by a Poisson process, Z_k are outcomes of independent r.vs. exponentially distributed with an average equal to 1. The independence is usually ensured by separating out peak values from different wave storms. The type of residue probability distribution can be checked again by means of the Q-Q and P-P plots. Furthermore, the line which fits the residues, and has been plotted against the corresponding times T_k, must be parallel to the time axis at $y = 1$.

There is a last, but important, question to be discussed. It concerns the value of the threshold u of a Poisson-GPD model. Once a value for such a parameter has been chosen, the model can be fitted to the data. We know, according to the theory, that this quantity should be close to the right end point of the actual probability distribution function of the data. In practice, the only way to identify an optimal threshold—and then the other parameters of the model—is by making several attempts.

9.3 Results of the Extreme-Event Analysis

Among the available tools for a statistical modeling of extreme events, we have chosen the software R with its packages of functions. Although we are most interested in the extreme events, R, with packages of functions, makes it possible to perform almost any kind of statistical analysis and modeling.

Roughly speaking, R has been developed to replicate S-Plus. The main advantage of R is that it is free: it can be downloaded from the R-project web site, http://www.r-project.org/. It is an open source software, so users can easily control all the available elements and modify them according to their specific needs. The documentation is often updated, and examples and further references are given. In R, big data sets are handled with good performance. Missing values can be marked as not available (NA), so that the analysis can be consistently performed.

Packages such as *evir* and *ismev* have been designed and developed for the analysis of extreme events. These packages have already been widely employed in scientific and technical researches in different fields. In particular, we have exploited the functions *gpd, plot.gpd* and *gpd.fitrange, gpd.fit, gdp.diag*. For their descriptions, we suggest reading the original documentation (http://www.r-project.org/).

In the next subsection we show, in detail, the results for the case of Alghero's site, which has the most intense SWH time series among the Italian sites of wave measurements. Later, we also show a summary of the results regarding the other sites.

9.3.1 Results: Alghero's Site

Alghero's site has a single principal sector of wave directions. As it happens, the seasonal components can be ignored because their inclusion in the analysis has not actually improved the quality of the results. This is not valid for all the sites, however, so we report it whenever a seasonal characterization has improved the analysis significantly.

Table 9.5 shows the values of the parameters which give a reasonable quality of the corresponding fit.

Let us observe that ξ, which is the shape parameter of the GPD, is negative. This means that the probability distribution function has a finite right end point. In other words, wave storms cannot have any peak values. Moreover, the obtained value is above -0.5, which ensures the convergence of the fit procedure and the asymptotic normal distribution of the parameter estimations.

In this case, 80 peaks have been selected from clusters of exceedances over a threshold fixed at 5.3 m, this being the 99.8% quantile of the data. Two different clusters of exceedances are defined as separated if between them there are three days

Table 9.5. Alghero's parameters.

Alghero	No season Single sector
Threshold (m)	5.3
N. of peaks	80
Quantile	99.8%
ξ	-0.375
ξ s.d.	0.091
σ	163
σ s.d.	23
λ	0.001956
H_s max.(m)	9.1

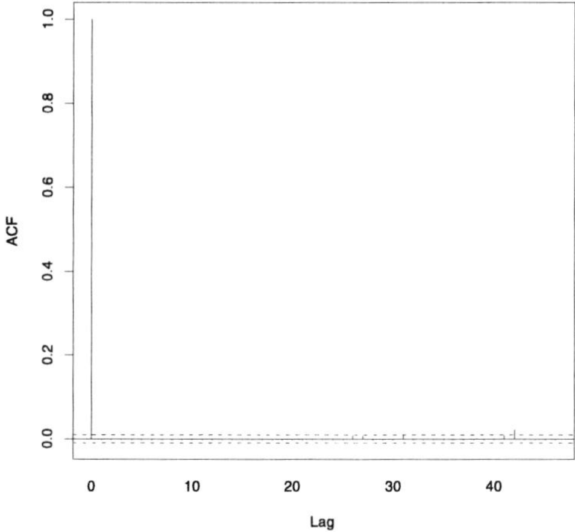

Fig. 9.2. Autocorrelation of the Alghero peaks.

of data below the given threshold. In other words, different storms are separated by calm periods. Using this scheme, it is reasonable to say that the selected peaks are outcomes of independent random variables, if they are not correlated. Fig. 9.2 strongly supports such a conclusion: in fact, the autocorrelation is equal to zero after the first lag within a confidence level of 95%.

Fig. 9.3 reports the P-P, Q-Q, return level and density plots of the r.vs. $T_k - T_{k-1}$.

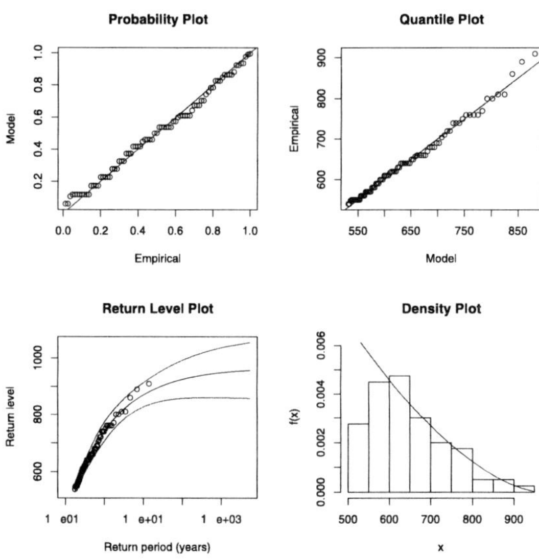

Fig. 9.3. Alghero site. From above, left to below, right: PP-plot; QQ-plot; return level plot and density distribution function plot.

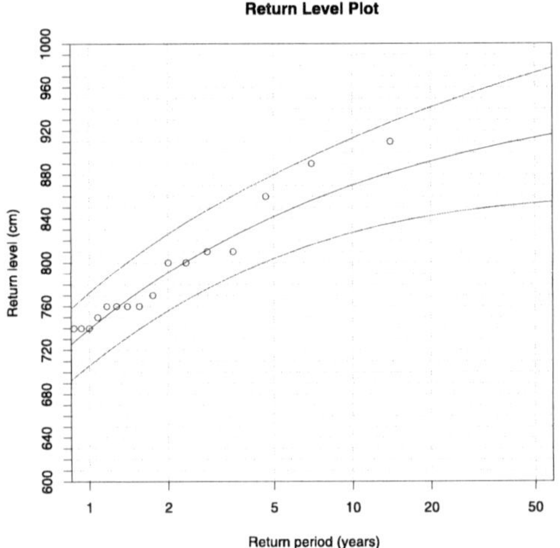

Fig. 9.4. Alghero return levels up to 50 years.

Here, we can see that the model fitting is in tune with the empirical estimates within a confidence level of 95%.

We should note here that the *return level* is defined as the level predicted to be exceeded in exactly one cluster of exceedances in m years. Moreover, a level plot can be considered the most important information for designing structures for the sea environment. A confidence interval for such an evaluation has to be provided. In our case, from the 14 years of data, we extrapolate a return level of up to 50 years (see Fig. 9.4).

Fig. 9.5 displays the scatter and Q-Q plots to test the residue properties, again ensuring the accuracy of the fit.

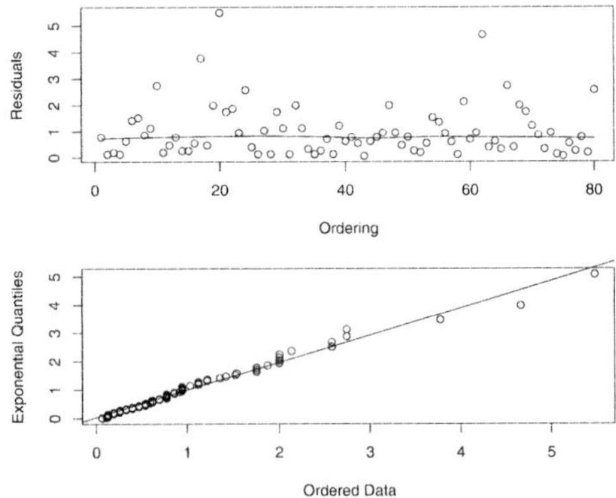

Fig. 9.5. Alghero case. Above: residues scatter plot; below QQ-plot against the exponential type.

9.3.2 Results for Other Sites

Here, Poisson-GPD model parameters for the sites of Catania, Crotone, La Spezia, Mazara, Monopoli, Pescara and Ponza are reported. Depending on the site, it was necessary to consider different principal sectors of wave directions and seasonal feaures. As in the case of Alghero, the seasonal analysis did not improve the results for the site of Pescara and for a sector of Crotone and Mazara.

A common result is that the shape parameter of the GPD pdf (probability distribution function) of all these stations, ξ, is always negative. This means that the actual pdfs are of the same type; the most important fact is that they have finite right end points. This is understandable. In fact, the energy of a storm is expected to be limited even if it is very intense. Moreover, the shape parameters are never less than −0.5, which ensures the convergence of the fitting procedures and the asymptotic normal distribution of the parameter estimates. Tables 9.6, 9.7, 9.8, 9.9, 9.10, 9.11 and 9.12 show the model parameters for each case.

9.4 Neural Networks and Sea Storm Reconstructions

In this application we focus on the reconstruction of SWH (H_s) data and on the effect of these data on the extreme-event statistics [19].

Here we consider the reconstruction of H_s for La Spezia,, a city in the northwest of Italy, with an important harbor. This site undergoes intense extreme events. In order to obtain such reconstructions, we introduce a system of NNs. The aim is to obtain uniform performances with respect to the possible values of the missing data. For this reason we have designed the system concerning the characteristics of the time series and the distribution of the corresponding data. In particular, we introduce some elements specialized for each season and both high and low values. In this way we will try to obtain a good reconstruction of any sea wave storm as well as the calm

Table 9.6. Poisson-GPD model parameters for Catania site.

Catania	Winter Single sector	Summer Single sector
Threshold (m)	2.2	1.7
N. of peaks	62	60
Quantile	99.6%	99.8%
ξ	−0.320	−0.202
ξ s.d.	0.111	0.112
σ	137	94
σ s.d.	22	16
λ	0.00442	0.00223
H_S max.(m)	5.7	4.7

Table 9.7. Poisson-GPD model parameters for Crotone site.

Crotone	Winter I sector	Summer I sector	No season II sector
Threshold (m)	1.8	1.7	2.3
N. of peaks	83	40	68
Quantile	99.5%	99.8%	99.8%
ξ	−0.434	−0.451	−0.239
ξ s.d.	0.127	0.129	0.090
σ	190	139	76
σ s.d.	31	27	11
λ	0.00474	0.00171	0.00166
H_s max.(m)	5.6	4.5	4.7

Table 9.8. Poisson-GPD model parameters for La Spezia site.

La Spezia	Winter Single sector	Summer Single sector
Threshold (m)	2.8	2.1
N. of peaks	69	106
Quantile	99.6%	99.6%
ξ	−0.262	−0.411
ξ s.d.	0.094	0.092
σ	155	128
Σ s.d.	23	16
λ	0.00410	0.00440
H_s max.(m)	7.4	4.9

Table 9.9. Poisson-GPD model parameters for Mazara'site.

Mazara	Winter II sector	Summer II sector	No season I sector
Threshold (m)	3.0	2.5	2.3
N. of peaks	69	65	73
Quantile	99.5%	99.8%	99.8%
ξ	−0.368	−0.463	-0.308
ξ s.d.	0.137	0.106	0.83
σ	131	108	106
σ s.d.	23	16	15
λ	0.00492	0.00241	0.00178
H_s max.(m)	5.9	4.6	5.2

Table 9.10. Poisson-GPD model parameters for Monopoli site.

Monopoli	Winter I sector	Summer I sector	Winter II sector	Summer II sector
Threshold (m)	1.7	1.6	2.0	1.7
N. of peaks	74	35	66	45
Quantile	99.5%	99.9%	99.6%	99.8%
ξ	−0.235	−0.309	−0.408	−0.283
ξ s.d.	0.096	0.144	0.120	0.115
σ	109	83	113	73
σ s.d.	16	18	19	13
λ	0.00480	0.00137	0.00428	0.00177
H_s max.(m)	5.1	3.7	4.4	3.7

Table 9.11. Poisson-GPD model parameters for Pescara site.

Pescara	No season Single sector
Threshold (m)	2.7
N. of peaks	73
Quantile	99.8%
ξ	−0.367
ξ s.d.	0.083
σ	124
σ s.d.	17
λ	0.00178
H_s max.(m)	5.7

Table 9.12. Poisson-GPD model parameters for Ponza site.

Ponza	Winter I sector	Summer I sector	Winter II sector	Summer II sector
Threshold (m)	2.0	1.7	2.7	2.5
N. of peaks	55	76	62	75
Quantile	99.6%	99.7%	99.5%	99.7%
ξ	−0.342	−0.288	−0.288	−0.274
ξ s.d.	0.134	0.107	0.088	0.080
σ	149	85	154	101
σ s.d.	26	13	23	14
λ	0.00436	0.00265	0.00492	0.00265
H_s max.(m)	5.5	3.9	7.1	5.5

periods. Afterwards, we shall evaluate in some detail how the insertion of the NN outputs affects the statistics for extreme events.

We remark that it is important to mantain the solution of feeding the system of NNs with the recorded data at nearby buoys. With this approach, data already reconstructed in previous steps are not used further, thus avoiding the enhancement of the error associated with filling in a long period of missing data [49].

Section 9.4.1 describes the peculiarities of our data set, such as seasonal variations and distribution of the data. In Section 9.4.2 we describe the system and the architecture of each element. The results are evaluated in section 9.4.3, where we focus on the reconstruction of extreme events. We carry out the extreme-event statistics using the peak-over-threshold (POT) method ([8], [17], [40], [41], [50], Chapter 6) to study the effect of the replacement of actual data by neural outputs on the estimation of the generalized extreme-value (GEV) distribution parameters. Finally, in order to further improve the reconstruction of clusters of exceedances over a high threshold, we introduce another approach, as described in Section 9.4.4 along with the results.

9.4.1 Data Analysis

Our aim is to obtain a good reconstruction for any sea wave storm as well as for calm periods. The idea is to model the architecture of a system of NNs on the characteristics of the H_s time series such as seasonal variations and data distribution.

We perform the analysis of the data set of measurements taken every three hours from 1990 to 2000. For each year we calculate the monthly mean values. We choose the month as a time interval because in this way we can observe the seasonal component of the phenomenon, filtering out the higher frequency. The plot in Fig. 9.6 of these values clearly shows different behaviors. The data from May to August (5, 6, 7, 8 in the figure) are characterized by average H_s values lower than the averages evaluated in the rest of the year. We call these two periods, respectively, the Summer and Winter seasons. Since high H_s values represent only a small portion of the set (Table 9.13), a single NN trained on the entire set would be mostly adapted to low values of the data. So we have decided to use, for each season, two different NNs: one for each type of data.

We plan to base the data reconstruction on the measurements performed by the buoys at the Alghero and Ponza locations, which are the nearest to the La Spezia site,

Table 9.13. Distribution of H_s La Spezia time series.

H_S value of La Spezia	Number of data from 1990 to 1995	Number of data from 1996 to 2000
<100 cm	12275	9726
100 cm–200 cm	2936	2845
200 cm–400 cm	999	918
>400 cm	61	51

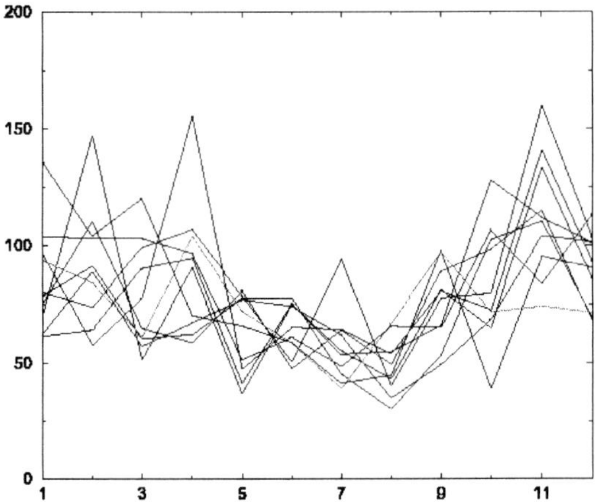

Fig. 9.6. Monthly H_s mean values from 1990 to 2000 for La Spezia's buoy. The y axis unit is in centimeters, the x axis represents months.

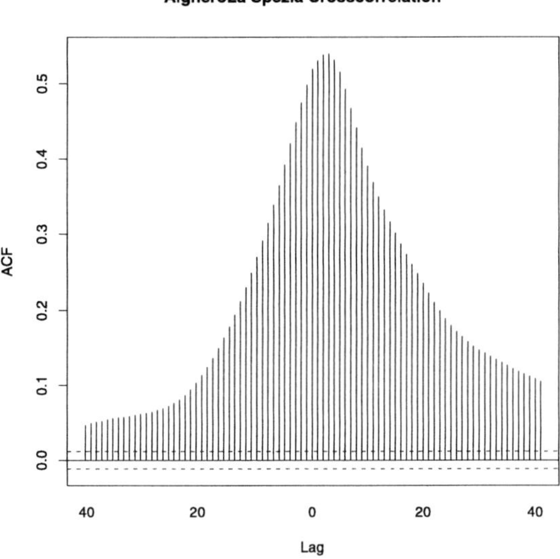

Fig. 9.7. Cross-correlation for La Spezia and Alghero H_s data from 1990 to 2000. The x axis shows the time lag (3-hour unit).

between the years 1999–2000. We call the buoys at Alghero and Ponza the reference buoys. Our idea is to use as input to the NNs the data that are the most correlated with the datum that we are going to reconstruct, at time t. For both the reference buoys these are the data at time $t + 3$ (i.e., time lag 3) and the ones around it (Fig. 9.7 and 9.8).

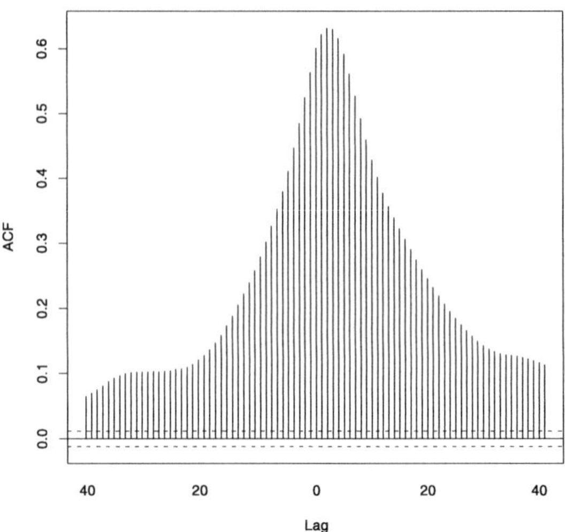

Fig. 9.8. Cross-correlation for La Spezia and Ponza H_s data from 1990 to 2000. The time lag (3-hour unit) is shown on the x axis.

Fig. 9.9 and 9.10 show the scatter-diagrams between the data of La Spezia at time t versus the data of the buoys of Alghero and Ponza at time $t+3$. It is clear that the data of the reference buoys cannot be used to obtain a reconstruction by a linear interpolation. In relation to La Spezia, Alghero's data turn out to be less similar than

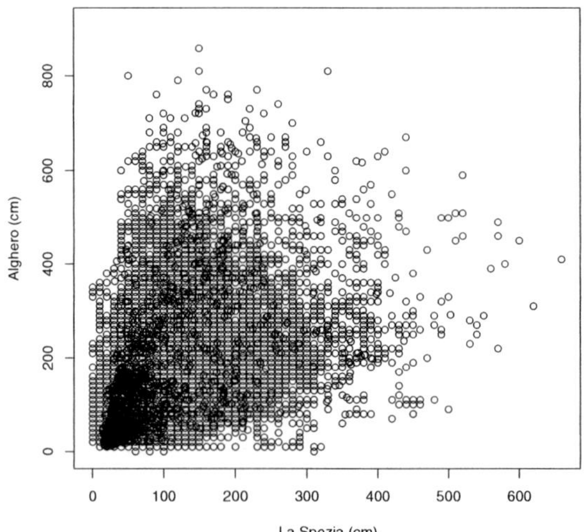

Fig. 9.9. Scatter-diagram for La Spezia-Alghero.

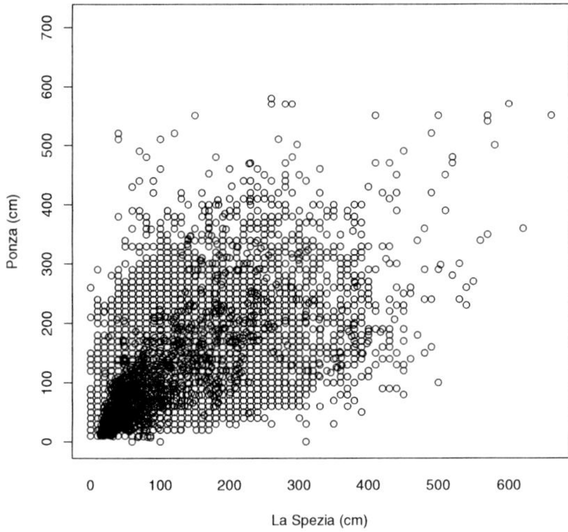

Fig. 9.10. Scatter-diagram for La Spezia-Ponza.

Ponza's. However, several preliminary tests have shown that feeding the NNs with Alghero's data improves the reconstruction performance.

9.4.2 The NN System

Let us suppose that the time series recorded by buoy A has gaps, while the two nearest (or most correlated) buoys B and C correctly perform the expected measurements. Then, we can try to recover the missing data by means of NNs which receive input data from both the B and C buoys (the reference buoys of A). This approach always uses available data from other buoys. It means that data already reconstructed in the previous steps are not used further. This has the advantage of successfully overcoming the enhancement of the reconstruction error that results when filling in gaps of long periods.

We introduce a system of NNs designed to work on the time series features discussed in the previous analysis in order to obtain good reconstructions of all the possible H_s values (Fig. 9.11). This will ensure a good reconstruction of any sea wave storm as well as calm periods.

The system actually has two subsystems: one for each season that was determined in section 9.4.1. Each subsystem consists in two NNs: one trained to reconstruct high H_s values, the other trained on low ones. During the reconstruction process, the two NNs in each corresponding subsystem are activated by the following rule: if the sum of the datum at time $t + 3$ of each reference buoys is larger than the sum of the seasonal mean of the B and C buoys, then the NN for high values is activated; otherwise the other one is used. Note that during this process the A's data are missing, so it is not possible to define an activation rule based on the data of the buoy A.

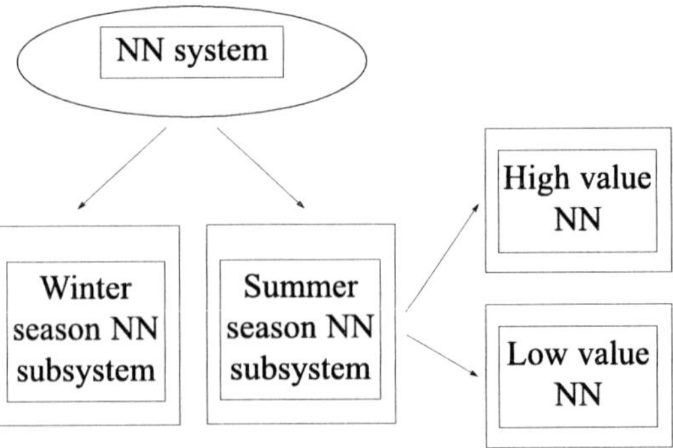

Fig. 9.11. Scheme of NNs system architecture. Each season NN subsystem consists of two NNs (High and Low value NN).

Fig. 9.12 shows the architecture of each element of the system. Each element is a two-layer feed forward perceptron (see Chapter 4). They have the same layer dimensions and transfer functions. Signals are sequentially forwarded from the input to the hidden layer, and from the hidden to the output layer. Neurons of the input layer have a linear transfer function, while the others have a sigmoidal one. Using sigmoid transfer functions ensures that an NN is equivalent to a complete system of functions (see Chapter 5). The output layer has one neuron, which gives the reconstructed datum at time t. The hidden layer has 12 neurons. The input layer has 8 neurons that receive the information from the reference buoys, four neurons for each buoy. In the case of La Spezia the first three inputs, for each reference buoy, are H_s at times $t+2$, $t+3$ and $t+4$; the fourth input is the average of the wave direction D_m for the same times. We note that these times correspond to the most correlated H_s of the references buoys to La Spezia H_s at time t. Moreover, attempts with other values have given worse results.

During the learning phase we minimize the following cost function (i.e., the learning error):

$$LE = \frac{1}{N} \sum_{t=1}^{N} p(H_s^A(t))(H_s^A(t) - O^A(t))^2, \tag{9.2}$$

where $O^A(t)$ is the NN output, $H_s^A(t)$ is the H_s value for buoy A and $p(*)$ is a weight function that is equal to 2 for high H_s^A values (values greater than half of the maximum of the H_s^A in the learning set), otherwise it is equal to 1. Hence, in the learning phase, this function weights the high H_s^A values more.

We note that the condition of activation of the NNs specialized on high or low values is based on the H_s values of the reference buoys, while the weight function is applied on the H_s^A values. Therefore, there are two checks: the latter on the output (only during the learning process), and the former on the input. Essentially, all these conditions are necessary because high value data represent only a small portion of the data set, as shown in Table 9.13.

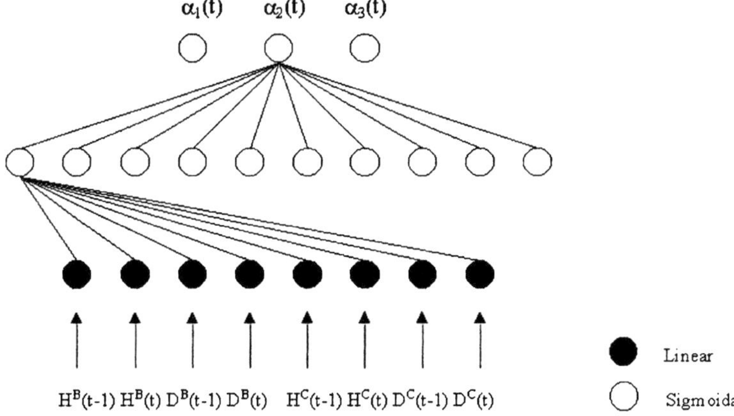

Fig. 9.12. Scheme of NN architecture. Neurons of the hidden layer are linked with all the neurons of the previous and of the next layers. The figure does not show all the connections but just a sample.

In such applications, cost functions usually have more than one local minimum. In this case, it is well known that the minimization of the cost function could not end up in the deepest minimum. In order to overcome such a problem, we use as a minimization algorithm the simulated annealing procedure with the Geman and Geman schedule. The procedure is a stochastic search, which make it possible to jump out of local minima reaching, asymptotically, the absolute minimum with probability 1 (see Chapter 4). For optimization of the NN performance, it is essential to find the right dimension of the hidden and input layers. Usually, this is accomplished using empirical attempts. Moreover, the generalization ability of the NN has to be tested on a different data set from the learning one (that is, the testing set). Indeed, this validation makes it possible to evaluate any possible over-fitting of the learning data.

9.4.3 Time-Series Reconstruction Results and Extreme-Event Statistic

We use data from January 1990 to December 1995 and data from January 1996 to December 2000 for the learning and the testing phases, respectively. To analyze the results, we calculate the mean absolute error (MAR), the mean signed error (MSE), the root mean square error (RMS) and the Correlation (Co) between the reconstructed and real data. We only show the results for the testing set because they are similar to the learning ones.

Table 9.14 shows that the performances of the NN system are quite good on average. In particular, the correlation between the actual and the reconstructed data is high and the MSE indicates a small underestimation.

From now on, we shall numerically evaluate how the injection of data provided by NNs modifies the main statistical properties of extreme events due to sea storms. This makes it possible to understand if reconstructed data can be used to fill the

Table 9.14. Summary of the NN system results: errors and correlation.

Testing 1996–2000	MAR	MSE	RMS	Co
	32 cm	−3 cm	49 cm	0.71

original time series without changing the true extreme-event statistics ([19]). We have used the POT method ([8], [17], [55]) to carry out this analysis. As explained in Chapter 6, this method consists of fitting the exceedances over a high threshold, u, with a two-dimensional Poisson process. This model works for a sequence of independent identically distributed random variables (i.i.d.r.vs.). The temporal independence can be obtained by selecting peaks of consecutive clusters of exceedances separated by a given interval of time during which the observations stay below the threshold. Regarding the stationary requirement, we consider the data distribution roughly stationary during one season and we choose the threshold high enough to obtain peaks mainly in the same season: the one we have indicated as Winter in Section 9.4.1. Since the peaks in the Winter season are higher than the others, using the same high threshold for the entire year guarantees that only the Winter peaks will be selected.

For the model fitting we use the POT function available with the package *evir* of R ([41]). By means of this procedure we estimate the parameters of GEV (9.3) ([50]):

$$H(x;\xi,\mu,\sigma)=\begin{cases}\exp\left\{-\left(1+\dfrac{\xi(x-\mu)}{\sigma}\right)_+^{-1/\xi}\right\} & \xi\neq 0\\ \exp\left\{-\exp\left\{-\left(\dfrac{x-\mu}{\sigma}\right)\right\}\right\} & \xi=0,\end{cases} \tag{9.3}$$

where $x > u$ and the notation $(\cdot)_+$ ensures an argument greater than zero. The shape parameter ξ, scale σ and location μ are the GEV parameters. Graphical analysis of the Q-Q plot and of the mean excess function plot makes it possible to evaluate the fitness of the approximations. These diagnostic procedures suggest that we set the threshold u at 310 cm, corresponding to the 98.5% quantile of the data, and separation periods of 3 days between consecutive clusters of exceedances.

We analyze how the estimates of the extreme-value parameters change by replacing the actual H_s values with the neural outputs. We consider a data set from 1990 to 2000, and then we replace from 1996 to 2000 year by year. So we execute 6 steps: at first we have 11 years of real data, and at the last step 6 years of actual data and 5 years of neural outputs (which have been obtained in the testing phase). On average, 80 peaks are present in each year. As is shown in Fig. 9.13, 9.14 and 9.15, the replacements cause a variation in the statistical parameters that does not show a change in the type of distribution (with respect to the sign of the shape parameter). Moreover, it can be observed that each estimation belongs to the 95% confidence intervals of the other steps. However, by inspection of the time series we find that the reconstructed data either have some artificial peaks or miss some real ones. Hence, even

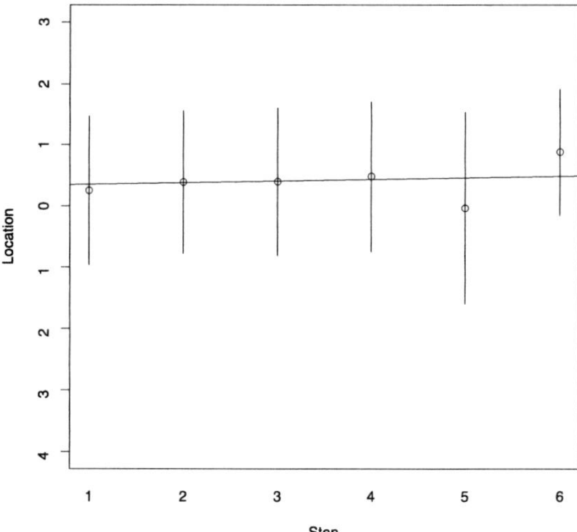

Fig. 9.13. Location parameter estimation for each step of the analysis on the testing set. The horizontal line is the interpolation of the parameters and the vertical lines are the bands of confidence.

if the global results are good, not all the storms or calm periods are properly reconstructed. We conclude that these poor reconstructions are, at least in part, balanced in the estimation of the GEV parameters.

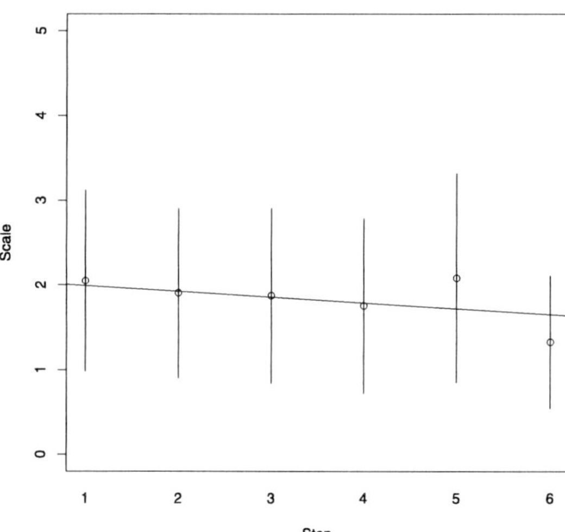

Fig. 9.14. Scale parameter estimation for each step of the analysis on the testing set. The horizontal line is the interpolation of the parameters and the vertical lines are the bands of confidence.

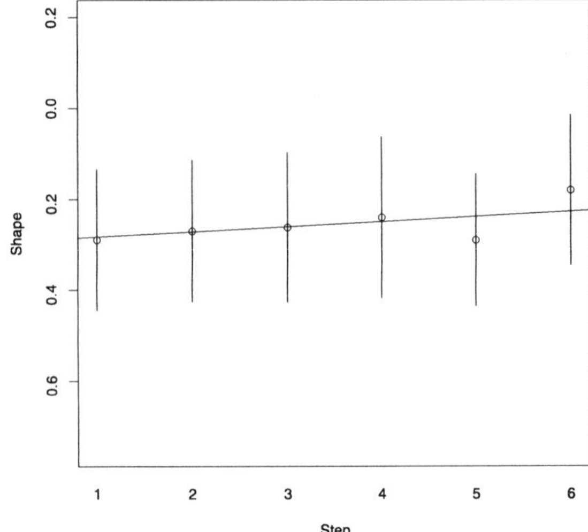

Fig. 9.15. Shape parameter estimation for each step of the analysis on the testing set. The horizontal line is the interpolation of the parameters and the vertical lines are the bands of confidence.

In order to further improve the reconstructions of storms, we introduce the approach described in the following section.

9.4.4 Extreme-Event Reconstruction

In this section, our aim is to improve the reconstruction of the clusters of exceedances over the threshold of 310 cm which are affected by gaps. The problems in the previous reconstruction were found by an analysis of the time series around each cluster of exceedances. This suggests that it is best to use a single NN locally trained around each event of interest. We get the best results using only the SWH as inputs of the NN, without including the directions of the waves. The generalization capability of this NN turns out to be optimized by executing a learning process on a set of 54 measurements (a week) around each cluster of exceedances over the chosen threshold. This NN works only in one cluster of exceedances. When considering another extreme event, we start with a new learning process on 54 measurements around it. The minimized *LE* is similar to 9.2, but the weight function is set to 1 under 250 cm, and to 15 above this level. In order to test the performance of this procedure, we introduced some fake gaps over the threshold of 310 cm for 18 clusters of exceedances. Fig. 9.16 shows the reconstruction of some of these events.

It is notable that about 44% of the considered events (8 on 18) have been reconstructed within an error of 10%. Moreover, due to the low correlation among exceedances over 310 cm of La Spezia with Ponza and Alghero (Fig. 9.17 and 9.18), this result could hardly be improved. We also remark that the reconstruction approach

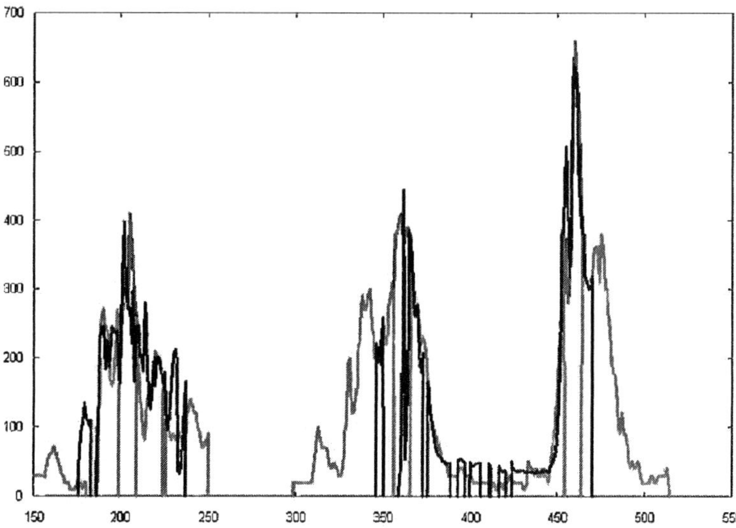

Fig. 9.16. Sample of local reconstruction. Different shaded areas show the real data of La Spezia, the same data but with the artificial gaps, and the reconstruction. When the NN cannot make a reconstruction due to a lack of input data it gives as output the key value -999. This justifies the vertical parts of the lines.

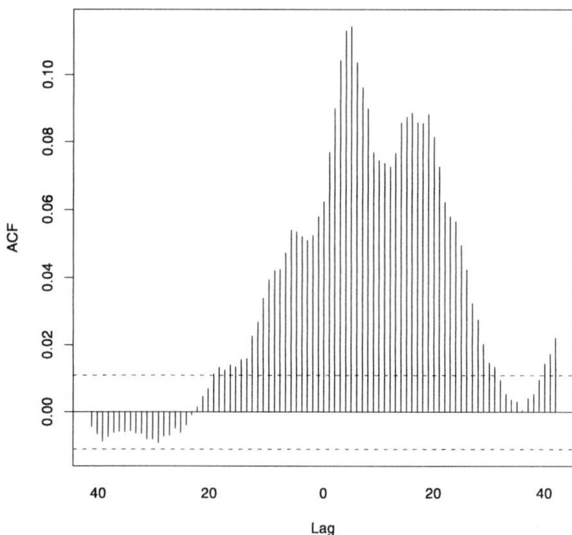

Fig. 9.17. Cross-correlation of exceedances over 310 cm of La Spezia with Alghero.

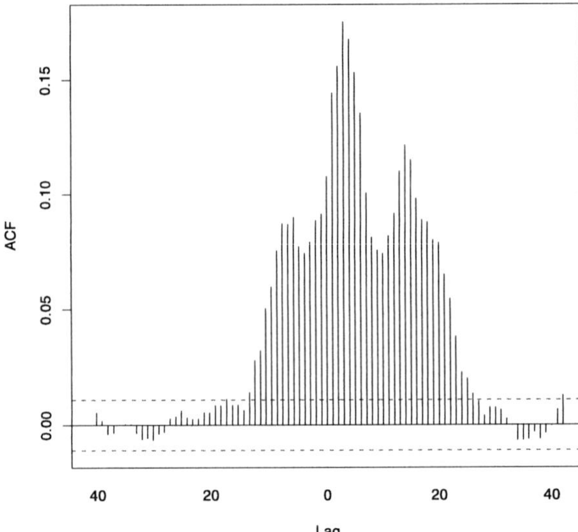

Fig. 9.18. Cross-correlation of exceedances over 310 cm of La Spezia with Ponza.

that has been adopted here can be successfully employed only for cases with a small number of gaps.

10

Generalization to Other Phenomena

In this chapter we deal with applications to situations different from the ones analyzed in the previous chapters in order to show the wide range of applications of the artificial neural network (ANN). Therefore, we apply the methods discussed in the previous chapters to the case of buoys moored off the the California coast, the postprocessing of temperatures for a certain gauge station in the south of Italy and to forecasting precipitation. In the first section we describe the structure of the network and of the optimal error chosen for the data reconstruction of the California coast. In the second section we show the application of this NN algorithm, proposed in Chapter 4 and applied in Chapter 7, and how it is able to model the SWH measured by the NOAA (National Oceanic and Atmospheric Association) National Data Buoy Center (NDBC-USA) near the California coast. In this application the NN uses as input the data of either one or more correlated data series from buoys located near San Francisco. The reconstruction of this time series is proposed by using different input data.

In the second section we analyze the capacity of the proposed NN algorithm to capture the relationship of the California coast SWH time series and to reconstruct gaps for missing data of long SWH time series with the same accuracy.

In the third section we show the application of ANN to a precipitation forecast. We did not follow the usual method because we wanted to explore the possibility of forecasts of precipitation using only the time-series data and NNs. We show that the main idea has been the proper selection of the input vector. In fact, we have chosen moving averages which smooth the local (in time) oscillations of the precipitation. As far as we know, this kind of quantity has not yet been considered as a useful variable in solving this problem.

In the last section we consider the corrections of a temperature forecast of the wave amplitude model (WAM) (explained in Chapter 3) using the values measured at a gauge station situated near the points of the WAM lattice. This postprocessing is done by means of ANNs and we compare it with the usual approach based on the Kalman filter [80]. This work is based on the postprocessing of the temperature forecast and uses a neural algorithm, which is a typical non-linear algorithm, that can capture the non-linear relationship among the values measured in a gauge station and the predicted values of the model. We discuss if it is better to use a linear algorithm for this correction, such as the Kalman filter, or a non-linear one like the ANN. From our comparison we conclude that the ANN method has a better performance than the linear corrections of the Kalman filter.

10.1 The Adaptive NN Algorithm and Its Learning Phase

The ANN is able to reconstruct gaps of a time series by using only data of different time series. For example, the relationship to be determined by the ANN may be the one between the SWH (H_s) of a station A and the measured values of the SWH, wave directions, wind directions (D) and wind intensities of other nearby stations. The method is based on the spatial-temporal overlapping of the wave climate data of a network of buoys, and on the analysis of the correlation between the SWH of station A and the data of each nearby station.

The algorithm makes a linear combination of a characteristic value of station A and of the SWH of the nearby stations, with the non-linear coefficients evaluated each time by the ANN. The data used may be from only one station (in this case the number of nearby stations N is equal to 1) or from more than one nearby station simultaneously ($N > 1$) ([49], [77], [78], [79]); in this case the ANN improves the performance but decreases the number of data that it may reconstruct. For the reconstruction it is better not to use data from many stations because in this case the ANN may reconstruct only the data recorded by all the N stations simultaneously. The two-layered ANN has L neurons in the input layer, M neurons in the hidden layer and Q neurons in the output layer, i.e., $Q = N + 1$.

Two error functions are used during the learning phase. The first is the mean absolute deviation (MAD) error (equivalent to the mean absolute error), which is commonly adopted in ANN applications:

$$LE_{MAD} = \frac{1}{P} \sum_{\mu=1}^{P} |H_{sA}^{\mu} - H_{so}^{\mu}|, \tag{10.1}$$

where P is the number of patterns belonging to the learning set, H_{sA}^{μ} is the SWH data that the NN has to approximate, the μth element of the learning set of P patterns, and H_{so}^{μ} is the output of the NN. The second function is a weighted error function that we define in order to calibrate the ANN for storm events reconstruction rather than for calm periods. It is given by

$$LE_{WL} = \frac{1}{P} \sum_{\mu=1}^{P} [C(H_{sA}^{\mu})|H_{sA}^{\mu} - H_{so}^{\mu}|], \tag{10.2}$$

where

$$C(H_{sA}^{\mu}) = C_i, \quad (i-1)m \le H_{sA}^{\mu} < im \tag{10.3}$$

is a step function and

$$C_i = \frac{P}{\text{cardinality}\{H^{\mu}_{sA}(i-1)m \leq H^{\mu}_{sA} < im\}}, \tag{10.4}$$

where $i = 1, \ldots, \text{int}(H_{\max})$, and $\text{int}(H_{\max})$ is the integer part of the maximum of SWH. The error function (10.2) is introduced because there are many low waves in the data series. Therefore, the ANN learns to reproduce these cases well and generalizes the data of higher sea waves badly. Let us note that with the *LE* given by (10.2) we emphasize the ability of the learning algorithm to minimize the error in the extreme-event reconstruction.

In the learning phase we have used simulated annealing (SA) (see Chapter 4) in order to get the absolute minimum. The structures of the chosen architecture of the NN and of the error are similar to the ones used in Chapter 7, confirming the validity of selecting the NN for dealing these problems.

10.2 The ANN Performance with One or More Time-Series Inputs

The NN algorithm is applied for the reconstruction of an SWH time series by using as input the data of correlated time series of SWH, wave mean period, wind intensity and direction obtained from nearby stations. A preliminary time-space analysis has been done to evaluate the embedding dimension of the problem. The best ANN performance is obtained by considering in the input the series of SWH and wind direction.

First the ability of the NN algorithm to reconstruct the SWH data of the 46012 NOAA buoy (San Francisco, California, USA), by processing the data of nearby buoys 46026 and 46042, has been tested ([77]; see also [49]). For this purpose, around $P = 8000$ patterns (one year of data), obtained from the data of 1999, have been introduced during the learning process. During this phase the following mean deviation (MD) error has been used:

$$LE_{MD} = \frac{1}{P}\sum_{\mu=1}^{P}(H^{\mu}_{sA} - H^{\mu}_{so}), \tag{10.5}$$

with its standard deviation and the mean absolute deviation (MAD) (10.1).

Three applications have been considered with time series with different cross-correlation values:

(i) input data are given by the time series of SWH and the wind directions of buoy 46026. In this case it is assumed that $N = 1$ and the input vector is given by

$$\bar{x}^{\mu} = (H^{\mu-1}_{46026}, H^{\mu}_{46026}, D^{\mu-1}_{46026}, D^{\mu}_{46026}), \tag{10.6}$$

with the SWH and the mean wind direction of the station 46026 at time m. We have $M = 8$ neurons in the hidden layer and $Q = 2$ neurons in the output layer;

Table 10.1. Learning phase.

Error	Case (i) (%)	Case (ii) (%)	Case (iii) (%)
MD error (m)	−0.04	−0.26	−0.06
Standard deviation (m)	0.28	0.36	0.22
MAD error (m)	0.28	0.35	0.17

(ii) the input vector is given by 46042 buoy data, with $N = 1, M = 8, Q = 2$; is therefore represented by

$$\bar{x}^{\mu} = (H_{46042}^{\mu-1}, H_{46042}^{\mu}, D_{46042}^{\mu-1}, D_{46042}^{\mu}); \tag{10.7}$$

(iii) the input vector is given by both 46026 and 46042 NOAA buoy data. Therefore, it is assumed that $N = 2$, $M = 8$ and $Q = 3$. The input vector is represented by

$$\bar{x}^{\mu} = (H_{46026}^{\mu-1}, H_{46026}^{\mu}, D_{46026}^{\mu-1}, D_{46026}^{\mu}, H_{46042}^{\mu-1}, H_{46042}^{\mu}, D_{46042}^{\mu-1}, D_{46042}^{\mu}). \tag{10.8}$$

The values of the errors during the learning phase for the applications (i), (ii) and (iii) are shown in Table 10.1. The testing error *TE* has been calculated as

$$TE = \frac{1}{P} \sum_{\mu=1}^{P} \frac{H_{sA}^{\mu} - H_{T}^{\mu}}{H_{T}^{\mu} H_{T}^{\mu}}, \tag{10.9}$$

and its value is shown in Table 10.2, where H_T^{μ} is the output of the ANN in the testing process. Let us note that the testing set is composed by data of years 1995–1998. Figure 10.1 shows the testing error *TE* (10.9) obtained for different ranges of SWHs. Note that the ANN algorithm also performs well for the higher sea waves.

The correlation coefficient between actual data and the output of the ANN is also calculated. As we can see from Table 10.2, we have the highest values for cases (i) and (iii). Figure 10.2 shows an actual time series recorded by buoy 46012 and the output of the ANN in cases (i), (ii) and (iii). As we can see, the ANN furnishes a good approximation of the SWH series, both during calm periods and during severe

Table 10.2. Testing phase.

Error	Case (i) (%)	Case (ii) (%)	Case (iii) (%)
Testing error TE (m)	−0.018	−0.081	−0.025
Correlation coefficient	0.95	0.89	0.95
Mean deviation (m)	−0.06	0.20	−0.07
Standard deviation (m)	0.28	0.35	0.25
Mean absolute deviation (m)	0.21	0.31	0.19

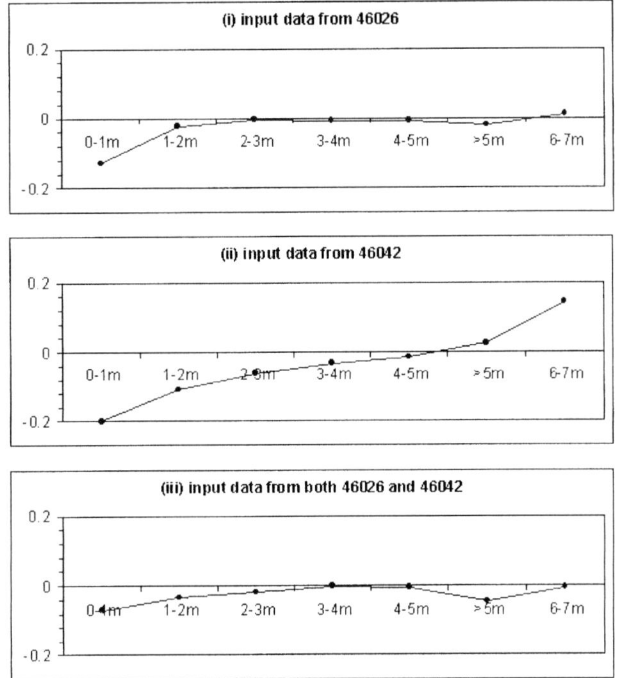

Fig. 10.1. Testing error *TE* for different ranges of significant wave height H.

storms, and the best performance is during application (iii). The ANN algorithm also performs well using only one station ($N = 1$), especially for the case (i) in which a very high value of the coefficient of correlation is obtained (see Table 10.2).

The ANN performance is slightly better for case (iii) with respect to case (i), but there is a reduction of the number of reconstructed data: the ANN is able to reconstruct only those data that are recorded simultaneously by the two stations 46026 and 46042. For example, in application (iii) the number of data of the testing set ($N = 20478$) is 8–15% smaller than in application (i) ($N = 22293$) and (ii) ($N = 23657$).

Figure 10.1 shows that the ANN algorithm may be applied for the reconstruction of gaps in the time series with good performance also by using the data of one buoy only as input (the higher the coefficient of correlation, the better the performance).

The ability of the ANN to model the extreme SWH, and therefore the sea storms, has finally been tested. Note that the modeling of extreme sea storm is very useful for engineering applications. For example, we may obtain the historical SWH time series at a fixed location, during the severest sea storm, which has damaged a buoy or a coastal structure.

Figure 10.3 shows data for a strong storm recorded off California in 1987, which is well reconstructed by the ANN. A sea storm is defined as a sequence of sea states

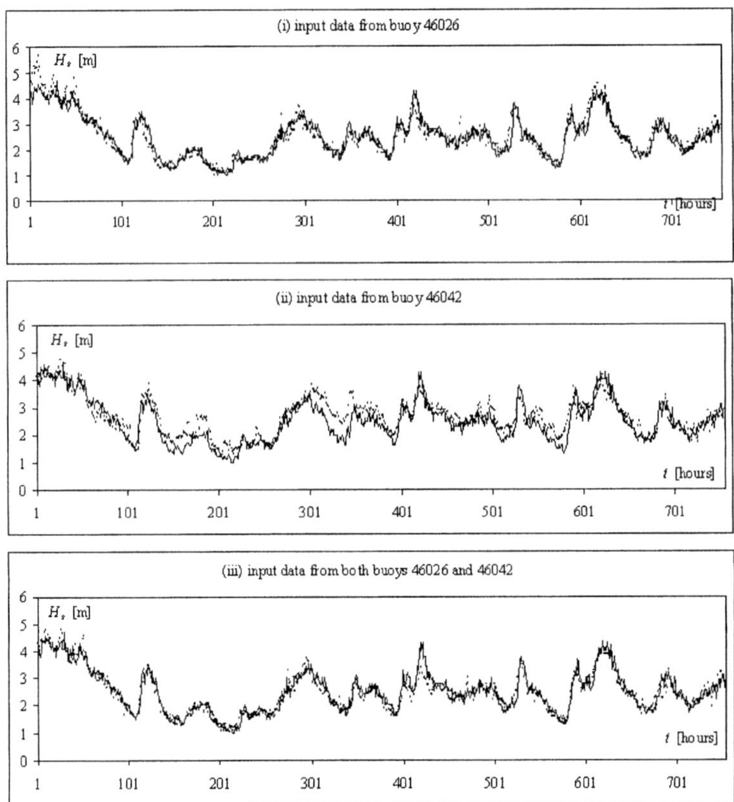

Fig. 10.2. Comparison between actual time series of SWH recorded by buoy 46012 starting from January 1995 (continuous lines) and the output of the ANN, for applications (i), (ii) and (iii).

in which the SWH Hs exceeds a fixed threshold (which is assumed equal to 1.5 times the mean SWH at the examined location) and does not fall below this threshold for a time span greater than 12 hours ([71]).

Note that the improvement of the ANNs ability to model sea storms (which was obtained by introducing the weighted error function) is very important for risk analysis. For example, the long-term statistics, which are very useful for the design of sea structures, enable us to obtain the SWH distribution at the examined location (see Fig. 10.1–10.3).

Fig. 10.4 shows the SWH distribution of buoy data, plotted in the Weibull paper (note that the Weibull distribution fits well the SWH distribution in many locations). Comparison with the output of the ANN shows a good agreement: the differences for the H_s values, for any fixed threshold of $P(H_s > H)$, are very small (14 years of testing set data have been used). A deeper treatment of risk analysis by using the output of the ANN was given in [78] (see also [79]).

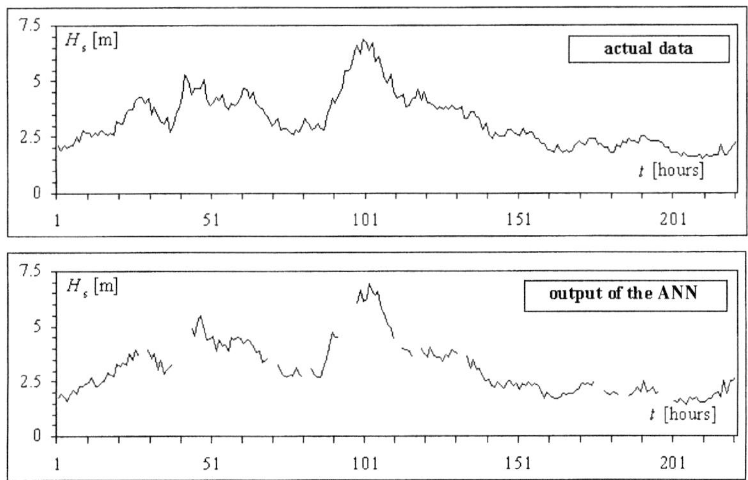

Fig. 10.3. A severe storm recorded by buoy 46012 and the corresponding data obtained by ANN, with input data given by both buoys 46026 and 46042.

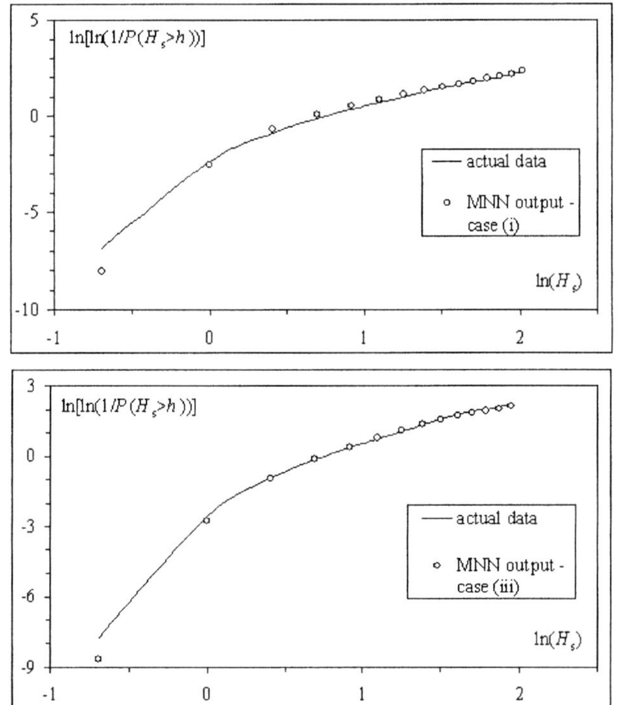

Fig. 10.4. The probability of exceedance of $P(Hs > H)$ of the SWH plotted in the Weibull paper.

10.3 NN and Precipitation Forecast

NNs are one of the most used algorithms for dealing with time-series analysis. The NN can find any relation inside a set of training data using a training procedure and a testing validation. In the particular case of NNs applied to the precipitation forecasts what we want to predict is the value of the millimeters of rain falling at time t using the data at previous times $t-1, t-2, \ldots, t-k$, with k suitably chosen. We have chosen this approach, which is not the usual one, because we believe that in the time series itself there is already enough information for the forecast if the series is long enough. We have used precipitation time series, where data were updated every 12 hours, in particular, those from the rain gauge of Torino/Caselle, in north Italy, from 1951 to 1958. In our simulations, we have used the average over 24 hours. We have used a two layer back-propagation NN with 6 input neurons, 18 hidden neurons and just one output neuron. We have chosen the Monte-Carlo (MC) method and simulated annealing (SA) as a minimization algorithm to adjust the internal parameters (Chapter 4). MC explores the space of the different configurations, while SA approaches the absolute minimum of the error. One important observation is that we have not introduced limitations on the range of data reproduced by the NN. This means that the NN could also predict negative precipitation amounts that are not physically reasonable. We have, in fact, noticed an improvement in the performance by allowing the NN to also predict negative values and by successively setting them to 0. On the basis of other research on the same problem, since we were interested in predicting the precipitation at time $t+1$, we have considered as input variables the first difference between the precipitation data at time t and $t-1$, along with the moving averages computed on a number of k days and evaluated on the first numbers of Fibonacci's series $k \in (3, 5, 8, 13)$. The moving averages are defined recursively as

$$MA_k(z_t) = \frac{1}{k}(z_t) + \frac{k-1}{k} MA_k(z_{t-1}).$$

The reason for this choice is that the Fibonacci numbers accurately represent the behavior of time series with long-range correlation. The moving averages are used for smoothing the local fluctuations of the time series. These quantities have been introduced in the science of economics to deal with the time series of the actions. We also considered, as input, many other different physical variables but our preliminary screening suggested that we concentrate on the pressure. We have carried out many simulations with different choices of the input variables; we report here only the most interesting cases, namely an NN with the first difference of precipitation data, four moving averages and the pressure. We are referring to a data set of 600 patterns, 450 for the learning phase and 150 for the testing one, but we prefer to show just the first 200 patterns to reach the best compromise between generality and visibility. The results are shown in Fig. 10.5 and 10.6. Another important aspect of the problem that we have considered is the treatment of all the events characterized by large precipitation values (extreme events). They are exceptional and rather rare. NN has been observed to be less effective in reproducing extreme events. This prompted us to use an appropriate error function to treat them, as we do in the case of buoy data

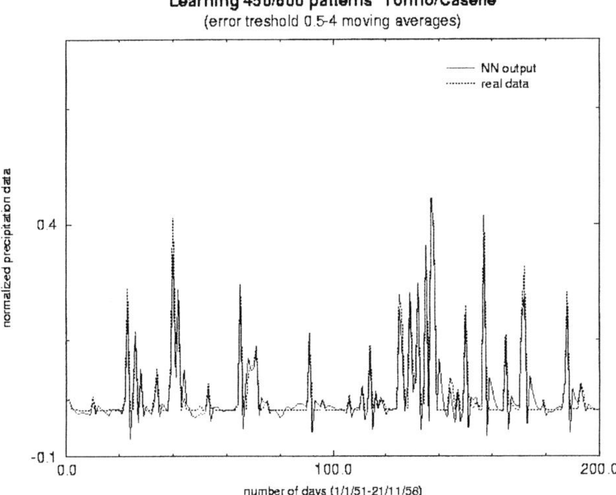

Fig. 10.5. Learning—NN with weight function.

and other cases examined in this book. The idea is to use, during the learning phase, an error function which generalizes the usual mean squared error by introducing a multiplication for a weight function p depending on the input data. This would cause the data values to be weighted in a different way. By using different probabilistic weights for different events, the NN could also learn and reproduce rare events. The

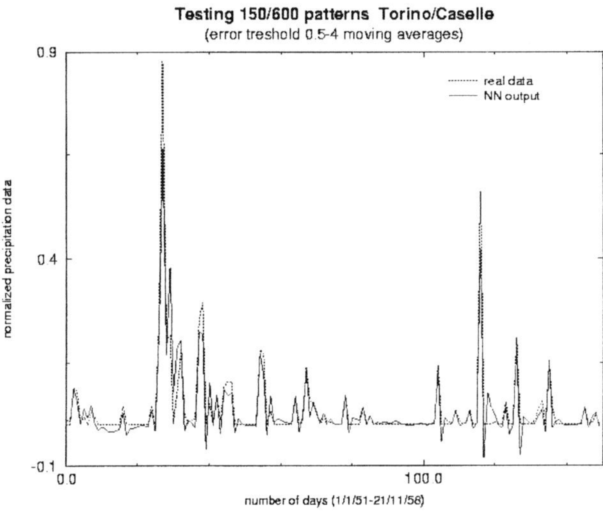

Fig. 10.6. Testing—NN with weight function.

same idea is applied in Chapter 7 for the SWH case and in the previous section for the SWH of the California coast. Thus, the weight function p in this case is

$$p(z^{\mu}) = p_{\min} + \frac{p_{\max} - p_{\min}}{1 + \exp(p_{th} - z^{\mu})/p_{st}},$$

where p_{th} is a threshold value and p_{st} a normalization factor. The introduction of this new error function has been of crucial importance because it has drastically improved the results given by the NN (see Fig. 10.5 and 10.6). After many different simulations we have understood that NN is a very useful tool for predicting precipitation amounts during 24 hours, while it is less good at correcting forecasts provided by models like the QBOLAM (a local area model implemented on a parallel machine called Quadrix). A possible explanation is that when we introduce as input variables the forecast provided by the meteorological model, if this forecast does not fit the original data very closely, the error in the NN learning set increases. In other words, it is difficult to carry out the postprocessing of the precipitations predicted by the QBOLAM due to the large errors of the relative forecast. Another postprocessing problem is discussed in the next section. In conclusion, we prefer to use the NN as a forecast tool using an appropriate error function and after the optimal choice of the variables. It might be possible to extend this result to the three-hour forecast, but it is unlikely to be accurate for periods of half an hour. The next step is to try to forecast precipitation events for these shorter times. That is, we should investigate the dynamics of convective cells of the atmosphere around the gauge region and during the considered precipitation events. Thus, the NN algorithm will be applied to forecast the dynamics of convective cells in the atmosphere and the evolution of the critical physical parameters connected with the onset of extreme precipitation events. This last approach could also be based on satellite data.

10.4 Postprocessing of Temperatures

Postprocessing is a general method of incorporating the information contained in measured data in the forecast of a model. We refer to the WAM (Chapter 3) or to other models of the same nature in this discussion. The model computes physical quantities such as the temperature (or precipitation, SWH, or SL) on a certain grid of points and performs forecasts at subsequent times on the basis of the values computed until a certain time. This forecast is sometimes wrong so one can try to correct the forecast using the data measured at a nearby station. The problem is that usually the location of the station does not coincide with the location of the points on the grid and that the relation between the value measured at the gauge station and the values on the grid computed by the model is typically non-linear, even if the points of the grid and the position of the gauge station are close together. The approach used so far has been to use a linear relationship defined by the Kalman filter ([80]). This is not very satisfactory because the connection is typically non-linear. We have also tried to use NN to obtain a better forecast ([80]). The most difficult part is, as usual, the choice of the input vector and the structure of the NN. Thus, we start from the forecasted temperatures for the ith day by the ECMWF at four grid points $T_i^1, T_i^2, T_i^3, T_i^4$ and we want to find a function $f(T_i^1, T_i^2, T_i^3, T_i^4)$ of these values connecting them with the measured temperature at the station To_i:

$$To_i = f(T_i^1, T_i^2, T_i^3, T_i^4). \tag{10.10}$$

So we look for an ANN with the output Tr_i as near as possible to the value of To_i. The input vector has been found after many attempts, the main difficulty arising from the fact that the correlation among the temperatures evaluated by the model has, as one would expect, very slow decay. We tried to drop the strong deterministic component of these quantities by making the first differences of the model temperatures $T_i^1 - T_{i-1}^1, \ldots, T_i^4 - T_{i-1}^4$. In order to obtain some interesting results the temperatures measured in the previous days were also included in the input vector $To_{i-1}, To_{i-2}, \ldots, To_{i-6}$. The best architecture found is an ANN with one input layer with 10 neurons, a hidden layer with 14 neurons and an output layer of only one neuron which gives the physical value. The learning error has been taken as the usual one:

$$LE = \frac{1}{P}\sum_{i=1}^{P} |To_i - Tr_i|, \tag{10.11}$$

but we were forced to take into account the non-stationary behavior of the time series. Again we have to deal with a problem of seasons. We should say that the time series is almost non-stationary because the average temperature value depends on the season. Thus, it is reasonable to find a best predictive model relative to the season. These seasons correspond in some way to the real seasons, but have been for the most part detected by inspection of the data. They are May 21–September 20 (first season); September 21–November 30 (second season); December 1–February 28 (third season); March 1–May 20 (fourth season). The almost stationary characteristic of the time series of the temperatures is included in this division. The learning error *LE* and the testing error *TE* for each network (or for each season) are of the same order of magnitude while the errors in the case of an NN are 1.4 (first season), 1.6 (second season), 1.4 (third season), 1.7 (fourth season). The corresponding errors for the Kalman filter application are: 1.8, 1.3, 1.5, 1.9. We find that the performance of the NN is better than that of the Kalman filter, which is the most powerful method used for the postprocessing technique. The non-linearity of the NN is essential; we verified that, if one uses linear transfer functions in the NN, the results of the *LE* and *TE* definitely become worse. In any case, we can conclude that the NN has been successful in giving a good approximation of the searched non-linear relation between the model forecast and the real values. This suggests that the same technique could also be used for other postprocessing problems, such as, for example, the SWH forecast of the ECMWF.

11

Conclusions

11.1 Summary

We have analyzed and explained many topics in this book. They differ very much but are all connected with the sea events and the algorithm selected for developing the analysis. The choice among the different algorithms has not been simple; we think that we have solved it in the optimal way, according to our taste and interests. The first principle used for collecting the various chapters has been to bring together all the theoretical and experimental facts concerning the sea time series and the phenomenology of the waves. Thus, the first three chapters have been devoted to an exposition of the main phenomenology (Chapters 1, 2) of the sea events as well as historical information. The measuring techniques are discussed and displayed in some detail so that the reader can feel the complexity of the process of collecting the data and also the practical and historical motivations which led to the development of such instruments. The waves and tides are distinguished and analyzed in detail. Chapter 3 covers the theoretical model currently used for forecasting sea waves. We discussed briefly the problems of the construction of the WAM which is used in Europe for the prediction of waves. We emphasize the fact that the actual derivation of this model disregards the free boundary nature of the problem. In Chapter 7 we compared the results of NN reconstruction with the output of the astronomical model for a check of the results and with the output of the WAM to check the order of magnitudes of the NN outputs. Chapters 4 and 5 give the main aspects and fundamentals of the theory and practice of NNs. Chapter 4 relies more on the probabilistic aspects of the NN theory and explains the various methods for preparing the application of a neural algorithm. Chapter 5 explains the NN from the point of view of approximating a given function by means of a set of complete functions, thus treating the classical formula for back-propagation as a linear combination of a complete set of functions. The two points of view are complementary, and we feel that it is enlightening to explain both. The theorems can be a bit hard to comprehend, but we tried to explain them without losing rigor. The numerous applications of Chapters 7, 8, 9 and 10 with their many variations and empirical insertions justify the need for a rigorous statement of the starting version of the algorithms. As the reader will see, the empirical procedures employed in these last chapters imply many changes; coefficients depending on the amplitude of the quantities can appear in the formula of the *LE* or the non-stationary time series can be divided in subsets where the stationarity is verified approximately. All these empirical choices are the result of many "trial and error attempts." In order to justify these heuristic changes one needs to be very confident with the initial formalism. An application of an NN is dependent on the data set and on the structure of the problem. A convenient choice of the NN is a complicated problem since NN algorithms belong to the class of disordered models, as we have shown in Chapter 4. In Chapter 6 we present the theory of another important aspect of the sea time series: the statistics of extreme events. The POT model is discussed and the case of dependent random v ariables is considered even if it is not applied to the data. This part of the theory is useful for the elaboration of an approach for dependent random variables. In Chapter 7 we show the reconstructions of time series made with the NN. The importance of this chapter is also that we show how, using a proper choice of NN, one can solve problems concerning large data sets; at least, this has been the case of our reconstructions. We also introduce the problem of non-stationarity in this chapter and we reduce it to the question of seasonal effects. This discussion starts in Chapter 7 and is developed in Chapter 9. The large amount of recovered data is a success of the method and these new data are available in the public domain. Another important issue is the reconstruction of extreme events. These are obviously the most important objects to deal with. Often an extreme event is a peak of a certain width, and some data of the peak may be missing. In this case we have shown in Chapter 9 that one can apply an NN specialized for the recovery of this special event, always using the values of the correlated stations as input. An important check performed in Chapter 9 involves the modification of the statistical properties when inserting data given by the NN. We checked that, on substituting real measures with data obtained with our neural reconstruction, the scale and shape parameters of the GPD do not change significatly. This fact makes the neural reconstructions more important. The application of a POT model to some data is also shown in Chapter 9. In Chapter 8 we have tried to do the same job with the approximation approach of the NN and ARIMA models. We determined that it is not a good approach. Finally, in Chapter 10 we show the power of the neural approach on the network of gauge stations near the California coast, again obtaining nice results. Chapter 10 is also devoted to other applications: forecast of precipitations and postprocessing of temperatures. So we have shown interesting possibilities for the NN to work on many classes of different measures. We think that this book strongly suggests that the NN approach is very useful in environmental science.

11.2 Open Problems

All the results obtained in this book and summarized in Section 11.1 bring us to a sequence of open problems to investigate.

- Postprocessing of WAM.

The results we obtained using the NN approach for the postprocessing of temperatures suggests applying the same procedure to the forecast of SWH operated by the WAM.

- An alternative model to WAM.
  The derivation of the WAM is not rigorous since, as we have stressed in Chapter 3, it ignores the free surface problem even if in the initial formulation this characteristic is present. In [61] a theory is constructed which includes this fundamental feature. The free boundary between atmosphere and ocean remains at all levels of this theory, and the governing system of partial differential equations) is solved using a non-local operator (pseudo differential operator) applied to the solution evaluated on the $z = 0$ surface. The equations are solved using the WKB approximation (Wentzell–Kramer–Brillouin approximation).
- Generalize extreme-value application to dependent random variables.
  The problem of having only independent random variables reduces long time series to a few data which are often not sufficient for supporting good statistical estimates. Perhaps using the theory of Chapter 6 it is possible to work with dependent random variables and include more data in the statistical estimates. This means getting lower values of the threshold u and an easier approach to the length of the time series, which could imply a strong non-stationarity effect.
- Reconstruction of extreme events.
  We have shown in Chapter 9 a good way of reconstructing the values of an important peak of SWH. We have to investigate further how this reconstruction improves the statistics.
- Application to coast evolution.
  One of the practical results of the recovery of missing data is that a larger data set is useful for obtaining information on the wave statistics. This information is used successfully for developing good models of coast evolution.

References

1. Makarynskyy, *Geophys. Res. Abstracts*, **5**, 04216, 2003.
2. M. Onorato, A. Osborne, R. Fedele, and M. Serio, Landau damping and coherent structures in narrow banded 1+1 deep water gravity waves, *Physical Review E*, **67**, 046305-1,6 2003.
3. K.B. Dysthe and K. Trulsen, Note on breather type solutions of the NLS as models for freak-waves, *Physica Scripta*, **82**, 48–52, 1999.
4. V.E. Zakharov, Direct and inverse cascades in the wind-driven sea, preprint, 2005.
5. E.N. Pelinovsky, Linear theory of the generation and instability of the wind waves for a weak wind, *Physics of Atmosphere and Oceans*, **14**, No. 11, 1167–1175, 1978 (in Russian).
6. E. Aarts and J. Korst, *Simulated Annealing and Boltzmann Machine*, John Wiley & Sons, Chichester, New York, Brisbane, Toronto, Singapore, 1990.
7. S.I. Amari, A universal theorem of learning curves, *Neural Networks*, **6**, 161–166, 1993.
8. C.W. Anderson, D.J. Carter, and P.D. Cotton, Wave climate variability and impact on offshore design extremes, Report prepared for Shell International, 26 March 2001.
9. G.A. Athanassoulis and Ch.N. Stefanakos, A nonstationary stochastic model for long-term time series of significant wave height, *J. of Geophysical Research*, **100**, No. 8, 16149–16162, 15 August 1995.
10. G.E.P. Box and G.M. Jenkins, *Time Series Analysis Forecasting and Control*, Revised edition, Holden Day, San Francisco, 1976.
11. C.L. Bretschneider, Revisions in wave forecasting: Deep and shallow water, *Proceedings of the 6th Conference on Coastal Engineering*, ASCE, Reston, VA, 1958.
12. C.K. Chui and X. Li, Approximation by ridge functions and neural networks with one hidden layer, *J. Approx. Theory*, **70**, 131–141, 1992.
13. C.K. Chui and X. Li, Realization of neural networks with one hidden layer, *Multivariate Approximation: From CAGD to Wavelets*, Ed. K. Jetter and F. Ultreras, World Scientific, Singapore, 77–89, 1993.
14. G. Cybenko, Approximation by superpositions of a sigmoidal function, *Math. Control, Signals, Syst.*, **2**, 303–314, 1989.
15. M. Darbyshire and L. Draper, Forecasting wind generated sea waves, *Engineering*, London, April 1963.
16. A.C. Davison and R.L. Smith, Models for exceedances over high thresholds, *J. R. Statist. Soc. B*, **52**, No. 3, 393–442, 1990.

17. P. Embrechts, C. Kluppelberg, and T. Mikosch, *Modelling Extremal Events for Insurance and Finance*, Springer, Berlin, 1991.
18. J. Feng, Generalization error of the simple perceptron, *Journal of Physics A: Mathematical and General*, **31**, No. 17, 4037–4048, 1998.
19. J.A. Ferreira and C. Guedes Soares, An application of the peaks over threshold method to predict extremes of significant wave height, *Journal of Offshore Mechanics and Arctic Engineering*, **120**, 165–176, 1998.
20. S. Geman and D. Geman, Stochastic relaxation, Gibbs distribution and Bayesian restoration of images, *IEEE Trans. on Pattern Analysis and Machine Intelligence*, **6**, 721–741, 1984.
21. G. Gyorgyi and N. Tishby, Statistical theory of learning a rule, *Neural Networks and Spin Glasses*, Ed. K. Thuemann and R. Koeberle, World Scientific Publishers, Singapore, 31–36, 1990.
22. K. Hasselmann, On the non linear energy transfer in a gavity waves spectrum, Part 1: General theory, *J. Fluid Mechanics*, **12**, 481, 1962.
23. K. Hasselmann, et al., Measurements of wind-wave growth and swell decay during the Joint North Sea Project (JONSWAP), *Deutsch Hydrogr. Z Suppl.*, **A8**(12), 95, 1973.
24. R. Hecht-Nielsen, Kolmogorov's mapping neural network existence theorem, in *Proceedings of the International Conference on Neural Networks III*, IEEE Press, New York, 11–13, 1987.
25. R. Hecht-Nielsen, *Neurocomputing*, Addison-Wesly, Reading, MA, 1990.
26. J. Herz, A. Krog, and R.G. Palmer, *Introduction to the Theory of Neural Computation*, Addison-Wesley, Reading, MA, 1993.
27. F. Takens, Detecting strange attractors in turbulunce, in: Dynamical Systems and Turbulunce-Lecture Notes in Mathematics, Vol. 898, D.A. Rand and L.S. Young, eds., Springer-Verlag, Berlin, 1981.
28. W.W. Hsieh and B. Tang, Applying neural network models to prediction and data analysis in meteorology and oceanography, *Bulletin of the American Meteorological Society*, 1998.
29. R.W. Katz, M.B. Parlange, and P. Naveau, Statistics of extremes in hydrology. *Advances in Water Resources*, **25**, 1287–1304, 2002.
30. C. Koch, *Biophysics of Computation, Information Process in Single Neurons*, Oxford University Press, London, 1999.
31. P. Janssen, The wave model, *Meteorological Training Course Lecture Series*, 2001.
32. J.C. Komen, L.Cavalieri, M. Donelan, K. Hasselmann, S. Hasselmann, and P.A.E.M. Janssen, *Dynamics and Modelling of Ocean Waves. A Variational Principle for a Fluid with a Free Surface*, Cambridge University Press, Cambridge, U.K., 1994.
33. R. Lama and S. Corsini, La Rete Mareografica Italiana, Istituto Poligrafico e Zecca dello Stato, Rome 2000.
34. M.R. Leadbetter, G. Lindgren, and H. Rootzen, *Extremes and Related Properties of Random Sequences and Processes*, Springer-Verlag, Heidelberg, Berlin, 1983.
35. B. Lenze, *On Multidimensional Lebesgue–Stieltjes Convolution Operators in Multivariate Approximation Theory, IV*, C.K. Chui, W. Schempp, and K. Zeller, eds., ISMN 90, Birkhäuser Verlag, Basel, 225–232, 1989.
36. B. Lenze, On constructive one-sided approximation of multivariate functions bounded variation, *Numer. Funct. Anal. and Optimiz.*, **11**, 55–83, 1990.
37. B. Lenze, A hyperbolic modulus of smoothness for multivariate function of bounded variation, *J. Approx. Theory and Appl.*, **7**(1), 1–15, 1991.
38. M. Leshno, V.Y. Lin, A. Pinkus, and S. Schoken, Multilayer feedforward networks with a nonpolynomial activation function can approximate any function, *Neural Networks*, **6**, 861–867, 1993.

39. J.C. Luke, A variational principle for a fluid with a free surface, *J. Fluid Mechanics*, **27**, 395–397, 1967.
40. A.J. McNeil and T. Saladin, Developing scenarios for future extreme losses using the POT model, in *Extremes and Integrated Risk Management*, P. Embrechts, PME, ed., RISK Books, London, available at www.math.ethz.ch/~mcneil, 2000.
41. A.J. McNeil, Estimating the tails of loss severity distributions using extreme value theory, *ASTIN Bulletin*, **27**, 117–137, 1997.
42. H.N. Mhaskar, Neural networks for optimal approximation of smooth and analytic functions, *Neural Computation*, **8**, 164–177, 1996.
43. H.N. Mhaskar and C.A. Micchelli, Approximation by superposition of sigmoidal and radial basis functions, *Adv. Appl. Math.*, **13**, 350–373, 1992.
44. H.N. Mhaskar and C.A. Micchelli, Degree of approximation by neural and traslation networks with a single hidden layer, *Adv. Appl. Math.*, **16**, 151–183, 1995.
45. MIAS (Marine information and Advisory Service) Catalogue of Wave Prediction Models 1Y83, MIAS Reference Publication No. 5. Institute of Oceanographic Sciences, Wormley, 1983.
46. J.W. Miles, On the generation of surface waves by shear flows, *J. Fluid Mechanics*, **3**, 185–204, 1957.
47. J.W. Miles, On Hamilton's principle for surface waves, *J. Fluid Mechanics*, **83**, 153, 1977.
48. M. Minsky and S. Papert, *Perceptrons*, MIT Press, Cambridge, MA, 1969.
49. S. Puca, B. Tirozzi, G. Arena, S. Corsini, and R. Inghilesi, A neural network approach to the problem of recovering lost data in a network of marine buoys, in *Proc. 11th ISOPE Conference*, Stavanger, Vol. III, pp. 620–623, 2001.
50. R.-D. Reiss and M. Thomas, *Statistical Analysis of Extreme Values with Application to Insurance, Finance, Hydrology and Other Fields*, Birkhäuser, 2001.
51. W.H. Munk and D.E. Cartwright, Tidal spectroscopy and prediction, *Philosophical Transactions, Royal Society, London Series A*, **259** (1105), 533–581, 1966.
52. E. Ateljevich, Delta Modeling Section 2000 Annual Report, Chapter 8, 21 Annual Progress Report, California Dept. of Water Resources, San Francisco Bay-Delta Evaluation Program, Ed. Michael Mierzwa, 2001.
53. S. Saks, *Theory of the Integral*, 2nd ed., Hafner Publishing Company, New York, 1937.
54. M. Shcherbina and B. Tirozzi Generalization and learning error for non linear perceptron, *Mathematical and Computer Models*, **35**, 259–271, 2002.
55. R.L. Smith, Threshold methods for sample extremes, in *Statistical Extremes and Applications*, J. Tiago de Oliveira, ed., Reidel, Dordrecht, 621–638, 1984.
56. R.L. Smith, Extreme value analysis of environmental time series: An application to trend detection in ground-level ozone, *Statistical Science*, **4**, No. 4, 367–393, 1989.
57. S.A. Solla, A Bayesian approach to learning in neural networks, *International Journal of Neural Systems*, **6**, 161–170, 1995.
58. S.A. Solla and E. Levin, Statistical mechanics of hypothesis evaluation, *Physical Review A*, **46**, 2124–2130, 1992.
59. B. Tirozzi, *Modelli Matematici di Reti Neurali*, CEDAM, Padova, 1995.
60. B. Tirozzi, The central limit theorem for significant wave height, 2002, preprint.
61. S.Yu. Dobrokhotov, B. Tirozzi, and T.Ya. Tudorovskiy, Asymptotic theory of water waves over nonuniform bottom, I. Multiscale reduced equations on the surface: from wind waves to tsunami, *Russian J. Math. Phys.*, **12**, No. 2, 2005.
62. A.C. Tuckwell, *Introduction to theoretical neurobiology*, Vol. 1, 2, Cambridge University Press, Cambridge, U.K., 1998.
63. F. Vallet, The Hebb rule for learning linerly separable Boolean functions: Learning and generalization, *Europhysics Letters*, **8** (8), 747–751, 1989.

64. V.N. Vapnik and Chervonenkis, On the uniform convergence of relative frequencies of events to their probabilities, *Theor. Prob. and Appl.*, **16**, No. 2, 264–280, 1971.
65. V.N. Vapnik, E. Levin, and Y. Lecun, Measuring the VC dimension of a learning machine, *Neural Computation*, **5**, 851–876, 1994.
66. V.N. Vapnik, *The Nature of Statistical Learning Theory*, Springer, Berlin, 1995.
67. B.V. Gnedenko, *The Theory of Probability*, Chelsea Publs. Co., New York, 1962.
68. M. Mezard, G. Parisi, and M. Virasoro, *Spin Glass and Beyond*, World Scientific, Singapore, 1987.
69. WAMDIG, The WAM model—A third generation ocean wave prediction model, *Journal of Physical Oceanography*, **18**, 1775–1810, 1988.
70. G.B. Whitham, *Linear and Non Linear Waves*, Wiley, New York, 1974.
71. P. Boccotti, *Wave Mechanics for Ocean Engineering*, Elsevier Science, Oxford, 2000.
72. R.S. Tsay and G.C. Tiao, Consistent estimates of autoregressive parameters and extended sample autocorrelation function for stationary and nonstationary ARMA models, *JASA*, **79** (385), 84–96, 1984.
73. F. Arena, On the prediction of extreme sea waves, in: *Environmental Sciences and Environmental Computing, vol. II*, pp. 1–50, 2004.
74. F. Arena and Barbaro, On the risk analysis in the Italian seas, CNR-GNDCI, Publ. No. 1965, BIOS, Italy, pp. 1–136, 1999.
75. R.G. Dean, Beach nourishment: Design principles, *Proc. Short Course attached to the 23rd Conf. Coastal Eng.*, 301–349, 1992.
76. F. Arena, On the beach nourishment stability in the Italian seas, *Proc. of 28° Convegno Nazionale di Idraulica e Costruzioni Idrauliche*, Potenza 2002, vol. IV, pp. 155–166, 2002.
77. S. Puca and B. Tirozzi, A neural algorithm for the reconstruction of space-time correlated series, *Seminarberichte aus dem Fachbereich Mathematik, Fern Universität, Hagen*, **74**, 81–91, 2003.
78. F. Arena and S. Puca, The reconstruction of SWH time series by using a neural network approach, *ASME Journal of Offshore Mechanics and Arctic Engineering*, **126**, No. 3, 213–219, 2004.
79. F. Arena, S. Puca, and B. Tirozzi, A new approach for the reconstruction of SWH time series, *Proc. of the 21st International Conference on Offshore Mechanics and Arctic Engineering, OMAE, 2002*, ASME (American Society of Mechanical Engineers), June 23–28, Oslo, Norway, pp. 1–8, 2002.
80. M. Casaioli, R. Mantovani, F. Proietti Scorzoni, S. Puca, A. Speranza, and and B. Tirozzi, Linear and nonlinear post-processing of numerically forecasted surface temperature, *Nonlinear Processes in Geophysics*, **10**, 373–383, 2003.

Index